Student Support Materials for AQA

A2 Physics

Unit 4: Fields and Further Mechanics

Author: Dave Kelly

William Collins's dream of knowledge for all began with the publication of his first book in 1819. A self-educated mill worker, he not only enriched millions of lives, but also founded a flourishing publishing house. Today, staying true to this spirit, Collins books are packed with inspiration, innovation and practical expertise. They place you at the centre of a world of possibility and give you exactly what you need to explore it.

Collins. Freedom to teach.

Published by Collins
An imprint of HarperCollinsPublishers
77-85 Fulham Palace Road
Hammersmith
London
W6 8JB

Browse the complete Collins catalogue at
www.collinseducation.com

10 9 8 7 6 5 4 3 2

ISBN-13 978-0-00-734385-0

British Library Cataloguing in Publication Data. A Catalogue record for this publication is available from the British Library.

Thanks to John Avison for his contribution to the previous editions.

Commissioned and Project Managed by Letitia Luff
Edited and proofread by Jane Roth
Typesetting by Hedgehog Publishing
Cover design by Angela English
Index by Jane Henley
Production by Leonie Kellman
Printed and bound by L.E.G.O. S.p.A. in Italy

Contents

3.4.1 Further mechanics

Momentum concepts

Examiners' Notes

Momentum can also be given in units of newton seconds (N s); this is exactly equivalent to $kg\,m\,s^{-1}$.

The **momentum** of a moving object is defined as its mass multiplied by its velocity. Momentum is a vector quantity, so its direction is the same as its velocity. There is no special name for the unit of momentum; its units are those of mass × velocity, $kg\,m\,s^{-1}$.

Definition

Momentum ($kg\,m\,s^{-1}$) *= mass (kg) × velocity* ($m\,s^{-1}$)   *or* $p = mv$

The momentum of a body is a measure of how difficult it is to stop it. A heavy lorry could have a mass of 40 tonnes. When it is travelling on the motorway at $25\,m\,s^{-1}$, the momentum is:

$p = 40 \times 10^3\,kg \times 25\,m\,s^{-1} = 1.0 \times 10^6\,kg\,m\,s^{-1}$

Compare this to the momentum of a person running at top speed:

$p = 80\,kg \times 10\,m\,s^{-1} = 800\,kg\,m\,s^{-1}$

Conservation of linear momentum

When there are no external forces acting on an object its linear momentum does not change. This same principle can be extended to a system of several objects. The **conservation of linear momentum** is an important principle in physics which can be stated as follows.

Definition

The total linear momentum of a system is constant provided that there is no external resultant force acting.

It isn't immediately obvious that this is true. A car standing at traffic lights has no linear momentum, yet a few seconds later it has gained momentum as it pulls away from the lights. This does not contravene the conservation of momentum, since if we are just thinking about the car, the friction of the road on the car is an external force and conservation of momentum does not apply. If we consider the car and the Earth as the 'system' there are no external forces acting, so conservation of momentum must apply. As the car gains momentum in a forward direction, the Earth gains momentum in the opposite direction.

Similarly, when you jump into the air, you are moving the Earth a little. The Earth has to acquire a momentum that is equal but opposite to your upwards momentum. Since your mass is much less than the Earth's, the Earth's velocity will be much smaller than yours.

Collisions

The conservation of momentum is often applied to situations where two objects collide. If we can ignore external forces, such as friction, the total momentum before the collision must be equal to the total momentum afterwards.

Suppose car A has a mass of 1500 kg and is moving at $20\,\text{m s}^{-1}$ when it collides into car B which has a mass of 1000 kg and is moving in the opposite direction at $10\,\text{m s}^{-1}$. After the collision the two cars stick together. We can use the conservation of momentum to calculate the velocity after the collision.

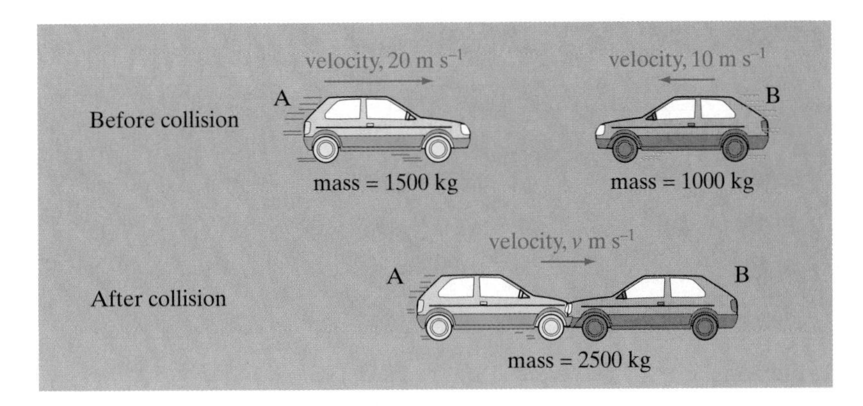

Fig 1
Conservation of momentum in a car collision

Momentum of car A = $1500\,\text{kg} \times 20\,\text{m s}^{-1} = 30\,000\,\text{kg m s}^{-1}$

Momentum of car B = $1000\,\text{kg} \times (-10\,\text{m s}^{-1}) = -10\,000\,\text{kg m s}^{-1}$

Before the collision the total momentum = $20\,000\,\text{kg m s}^{-1}$

After the collision the mass of the vehicles is 2500 kg and their velocity is $v\,\text{m s}^{-1}$.

The momentum after the collision = $2500 \times v$

If we can ignore any external forces the momentum before and after the collision must be the same:

$$20000 = 2500 \times v$$

$$v = \frac{20000}{2500} = 8\,\text{m s}^{-1}$$

Examiners' Notes

Remember that momentum is a vector quantity. If the momentum in one direction is positive, the momentum in the opposite direction must be negative.

Elastic and inelastic collisions

Collisions are classified as **elastic** or **inelastic**.

Definition

In an elastic collision there is no loss of kinetic energy.

If a collision is elastic, the total kinetic energy is the same before and after the collision. In an inelastic collision, kinetic energy is transferred to energy in different forms, such as heat. All collisions between everyday objects are inelastic – some energy is always transferred to other forms. Elastic collisions can take place between the molecules in a gas or between subatomic particles.

Elastic and inelastic collisions can be investigated in a school laboratory using gliders which run on an air track. The track has air blown through it

Essential Notes

Linear momentum is conserved in both elastic and inelastic collisions, provided there is no resultant external force.

so that the gliders rest on a cushion of air, eliminating friction between the glider and the track. If the air track is carefully levelled so that the glider moves horizontally, gravity will have no effect on the motion and we can say that there are no significant external forces acting on the glider. Light gates are used to time the glider's motion and a datalogger to calculate the velocity of the gliders so that the motion can be analysed (Fig 2).

Fig 2
Experimental apparatus to investigate elastic collisions

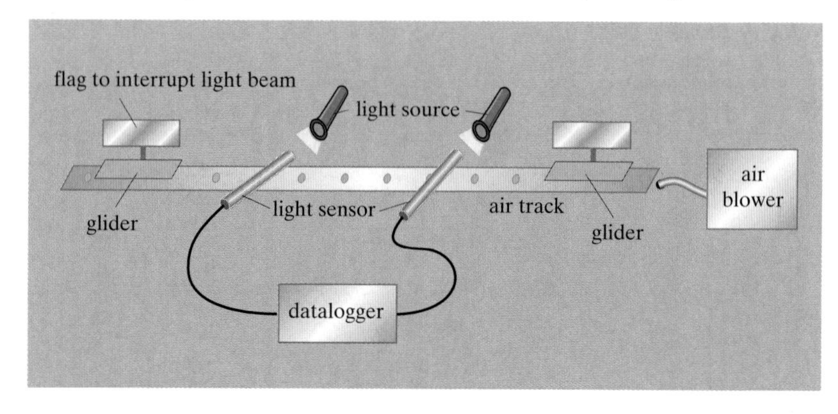

Explosions

The operation of a rocket engine can be explained by the conservation of momentum. The rocket engine propels hot gases out of the rear. The rocket must gain an equal momentum in the opposite direction (Fig 3).

Fig 3
Conservation of momentum in a rocket

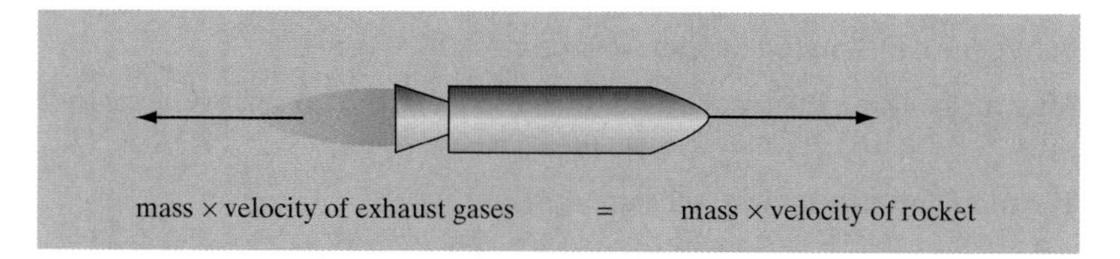

Example

A spacecraft of mass 20 000 kg which is at rest fires its rockets. The exhaust gases are expelled at a rate of $100\,kg\,s^{-1}$ and at a speed of $1000\,m\,s^{-1}$. If the rocket is fired for 10 seconds, what will the final velocity of the spacecraft be?

Answer

Total momentum before firing is zero. Therefore total momentum afterwards is also zero; the 'reverse' momentum of the exhaust gases must be balanced by the forward momentum of the rocket.

In 10 s the mass of exhaust gases is $100\,kg\,s^{-1} \times 10\,s = 1000\,kg$

Total momentum of gases $= mv = 1000\,kg \times 1000\,m\,s^{-1}$
$= 1\,000\,000\,kg\,m\,s^{-1}$

Momentum of spacecraft must also be $1\,000\,000 = 19\,000 \times v$, so $v = 53\,m\,s^{-1}$ (to 2 s.f.)

Note that the rocket's mass has decreased by 1000 kg

Examiners' Notes

It is a common misconception that rockets work by pushing against the Earth. This can't be true, because rockets work out in space where there is no external object to push against. Rocket engines push against the exhaust gases, which push back with an equal but opposite force on the rocket.

This is an example of Newton's Third Law (Unit 2).

The conservation of momentum also explains why a gun recoils when it fires a bullet. Before the gun is fired, the initial momentum of the system, that is the gun and the bullet, is zero. The final momentum of the system must also be zero. This means that any forward momentum acquired by the bullet must be balanced by the momentum of the gun recoiling in the opposite direction.

The same considerations apply to all types of explosions, such as a bomb exploding or the emission of an alpha or beta particle from a nucleus (Fig 4).

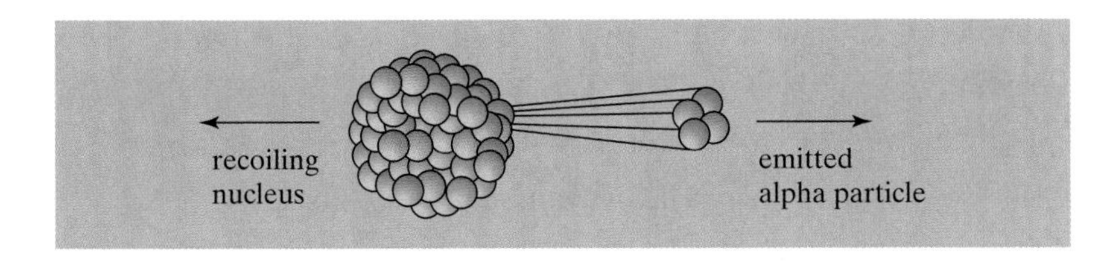

Fig 4
The momentum of the emitted alpha particle is equal but opposite to the momentum of the recoiling nucleus. Since the nucleus has a greater mass, it will recoil at a lower velocity than that of the alpha particle

Force and momentum

Newton's Second Law of motion defines a force as something that changes an object's momentum.

Definition

The rate of change of an object's linear momentum is directly proportional to the resultant external force. The change in momentum takes place in the direction of the force.

The force on an object, F, is proportional to its change in momentum divided by the time taken for the change:

$$F \propto \frac{\Delta p}{\Delta t} \quad \text{or} \quad F \propto \frac{\Delta(mv)}{\Delta t} \quad \text{or} \quad F = k\frac{\Delta(mv)}{\Delta t}$$

where k is a constant. In SI units the constant k is equal to one, by definition, and so the equation can be written

$$F = \frac{\Delta(mv)}{\Delta t}$$

The resultant force on an object is equal to its rate of change of momentum.

Essential Notes

When the mass m is constant, this equation becomes $F = m\Delta v/\Delta t = ma$ (see Unit 2).

Impulse

If we rearrange the equation above slightly, we obtain

$$F\,\Delta t = \Delta(mv)$$

and we see that the total change in momentum is equal to the product $F\,\Delta t$. This product, the force multiplied by the time over which it acts, is known as the **impulse**. It has units of newton seconds, N s.

Definition

***Impulse** is the magnitude of a force multiplied by the time for which it acts: the impulse of a force F that acts for a time Δt is F Δt.*

The impulse is useful for finding the change in momentum of an object.

Example

A golf club strikes a golf ball, of mass 45 g, with an average force of 5 kN. The contact between the club and the ball lasts for 0.5 ms. Find the velocity at which the golf ball leaves the club.

Answer

The impulse of the force is $F\Delta t = 5000 \times 0.5 \times 10^{-3} = 2.5\,\text{N s}$

This is equal to the change in momentum of the golf ball.

$\Delta(mv) = 2.5\,\text{N s}$

As the mass is constant,

$m\,\Delta v = 2.5\,\text{N s}$

so

$$\Delta v = \frac{2.5}{0.045} = 55.6\,\text{m s}^{-1}$$

This is the velocity at which the ball leaves the club, as initially it is at rest.

In many cases the force is not constant, but varies with time. The total impulse, and hence the change in momentum, is then given by the area under the force–time graph. For example, when a tennis racket hits a tennis ball, both objects are squashed. The ball and the racket act like springs. The force is initially small, increases to a peak value and then decreases again as the racket and ball return to their original shapes (Fig 5).

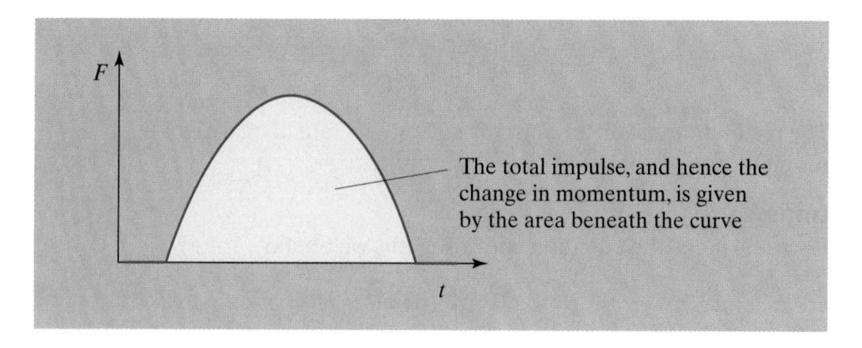

Fig 5
The force between a tennis racket and ball during a shot, against time

Water jets

In some cases, such as rockets and jets, mass cannot be treated as constant. It is then necessary to think of force as the rate of change of momentum, and to apply Newton's Second Law in the form:

$$F = \frac{\Delta(mv)}{\Delta t}$$

Example

A pressure washer can eject $1.5 \times 10^{-2}\,\text{m}^3$ of water per minute at a speed of $52\,\text{m s}^{-1}$. The water jet hits a wall and comes to rest. Find the force exerted by the jet on the wall. (Density of water $= 1000\,\text{kg m}^{-3}$)

Answer

In 1 second, the pressure washer ejects $\frac{1.5\times10^{-2}}{60} = 2.5 \times 10^{-4}\,\text{m}^3$ litres of water.

This is a mass of $2.5 \times 10^{-4} \times 1000 = 0.25\,\text{kg s}^{-1}$.

The water is moving at $52\,\text{m s}^{-1}$ so the momentum lost by the water in 1 second is:

$p = mv = 0.25 \times 52 = 13\,\text{kg m s}^{-1}$

Since force in newtons is equal to the change of momentum in 1 second, the force is 13 N.

Circular motion

Angular measure

At AS level you studied the equations of linear motion. Movement in a straight line can be defined by the time taken, t, the displacement, s, the acceleration, a, and the initial and final velocities, u and v. For an object moving in a circular path the displacement may not be important, since after each full circle the displacement is zero. It is often more useful to consider the total angle, θ, that has been turned through.

The SI unit used to measure angles is the **radian**. The radian is defined using a circle. The angle in radians, θ, at the centre of a circle is the ratio of the arc length, s, to the radius of the arc, r:

$$\theta \text{ (in radians)} = \frac{\text{arc length}}{\text{radius of arc}} \qquad \theta = \frac{s}{r}$$

Definition

One radian is the angle subtended at the centre of a circle by an arc that is equal in length to the radius.

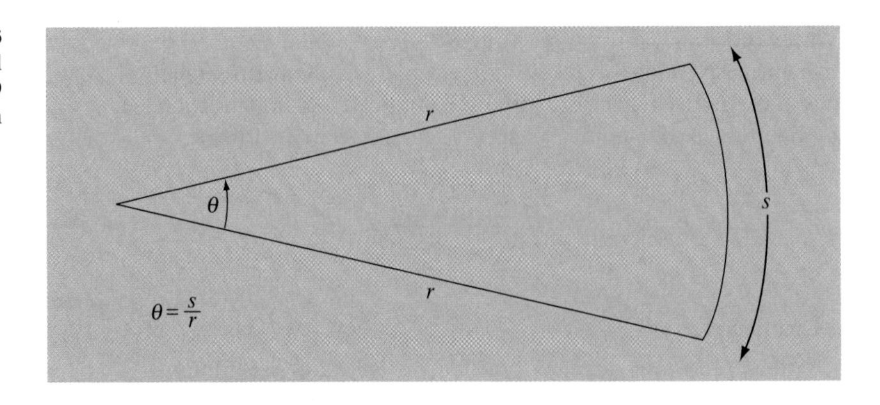

Fig 6
When the arc length is equal to the radius, the angle θ is equal to one radian

For a full circle the arc length is the circumference, which has a length of $2\pi r$, so:

$$\theta\,(\text{radians}) = \frac{2\pi r}{r} = 2\pi$$

In other words, there are 2π radians in a full circle. As there are 360° in a full circle, we can use this to convert radians to degrees (see Table 1).

Table 1
Conversions between radians and degrees

Radians	Degrees
2π	360°
1	$360/2\pi$ = 57.3°
$2\pi/360$ = 0.017	1°

Example

Calculate the angle in radians that the Earth spins through in one hour.

Answer

The Earth turns through a full circle, 2π radians, in approximately 24 hours. So in one hour:

$$\text{Angle } \theta = \frac{2\pi}{24} = 0.262\,\text{rad}$$

Rotational frequency and angular speed

The rate at which an object turns is often given in terms of the number of full circles that it completes in a given time. This is the **rotational frequency**, f, and it is typically quoted in revolutions per minute (rpm). The 'rev-counter' on a car is marked in rpm; the spin-speed of a washing machine is typically around 1000 rpm. The SI unit of rotational frequency is the hertz, Hz, which is the number of revolutions per second.

The **angular speed**, ω, is the angle turned through in one second, measured in radians per second, rad s^{-1}. Since there are 2π radians in one revolution, the angular speed is $2\pi \times$ the angular frequency, i.e.

$$\omega = 2\pi f$$

Linear speed

Even though an object is moving in a circle, we can still define its 'linear' speed as the distance covered per unit time. The speed, v, of an object moving in circular motion depends on the radius of the circle, r, as well as on the angular speed, ω. A child riding on a roundabout will have a low linear speed if they are sitting near the axis of rotation, but a much higher speed if they sit near the edge of the roundabout, even though the angular speed is the same in both cases.

Essential Notes

All points on a rotating object have the same angular speed, but different points have different linear speeds.

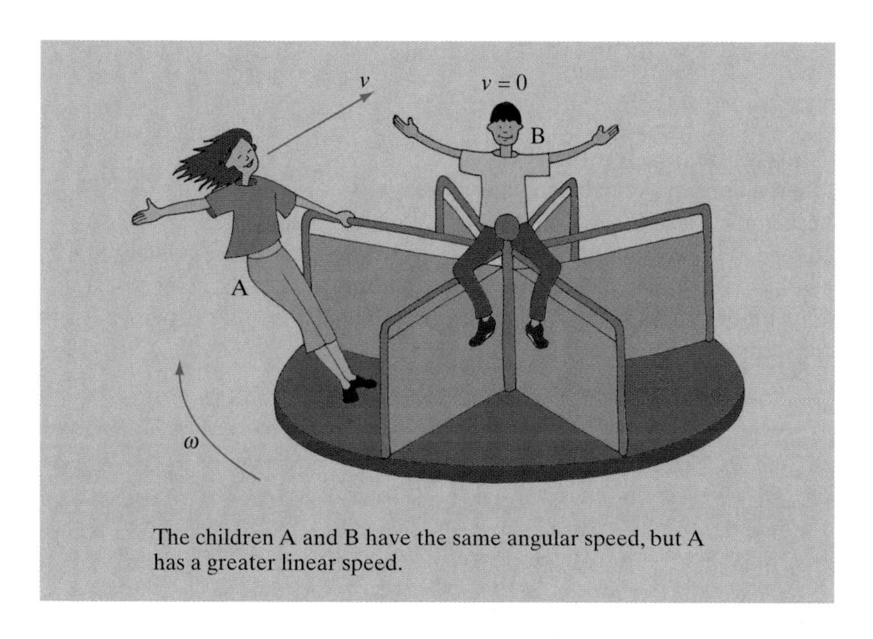

Fig 7
Speed on a roundabout

We can derive an equation connecting angular and linear speeds by considering the definition of an angle θ in radians:

Since $\theta = \dfrac{s}{r}$, if we consider the angle moved through in a certain time, t,

then $\dfrac{\theta}{t} = \dfrac{s}{rt}$

but $\dfrac{\theta}{t} = \omega$ and $\dfrac{s}{t} = v$,

so $\omega = \dfrac{v}{r}$

or

$$v = r\omega$$

Example

Find the rotational frequency and the angular velocity of the Earth due to its daily rotation about its axis. Find the speed of a point on the equator, given that the radius of the Earth is 6.4×10^6 m.

Answer

Rotational frequency is the number of revolutions per second:

$$f = \frac{1}{24 \times 60 \times 60} = 1.16 \times 10^{-5}\,\text{Hz}$$

The angular speed is given by

$$\omega = 2\pi f = 2\pi \times 1.16 \times 10^{-5} = 7.27 \times 10^{-5}\,\text{rad s}^{-1}$$

The speed of a point on the equator is

$$v = r\omega = 6.4 \times 10^6\,\text{m} \times 7.27 \times 10^{-5}\,\text{rad s}^{-1} = 465\,\text{m s}^{-1}$$

Essential Notes

Recall from Unit 2 the difference between **vector** and **scalar** quantities. Velocity is a vector quantity; it has a direction as well as a magnitude. Speed is a scalar quantity; it is defined by a magnitude only.

Centripetal acceleration

When an object's velocity changes, we say that it is accelerating. Acceleration is defined as the rate of change of velocity; this could be a change in the magnitude *or* a change in the direction of the velocity. Since an object moving in a circular path is continuously changing the direction of its motion, it must be continuously accelerating even if its speed is constant.

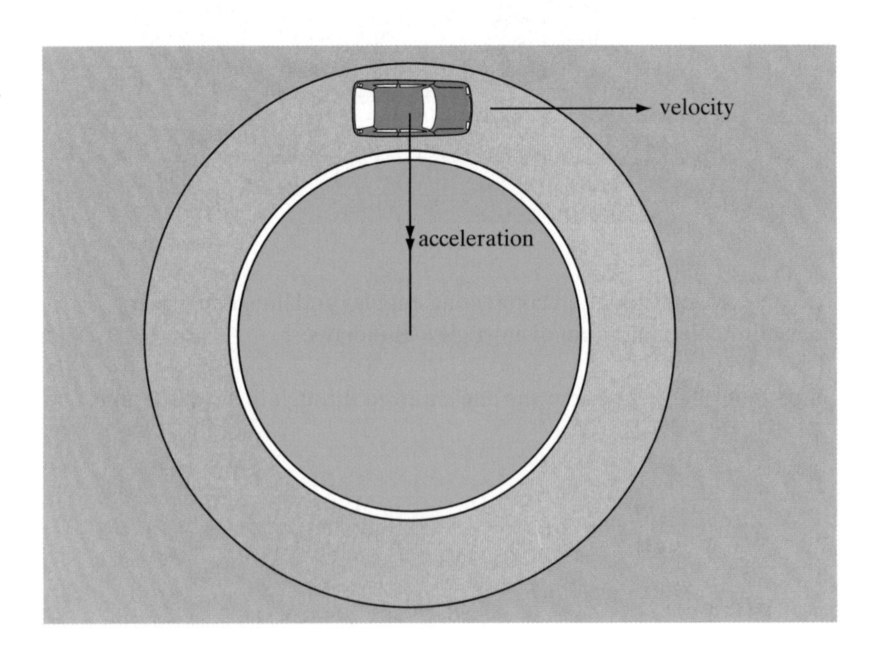

Fig 8
When a car goes round a roundabout or a corner, it is accelerating, even if it maintains a constant speed

Examiners' Notes

Objects that are travelling on a circular path may also change their speed: this results in an angular acceleration. However, this unit is limited to uniform circular motion, i.e. motion at constant speed.

The acceleration of an object in circular motion is always directed towards the centre of the circle. It is known as **centripetal acceleration**. The magnitude of this centripetal acceleration depends on how quickly the direction is changing. In a short time, Δt, the object moves through an angle $\Delta\theta$ and the velocity changes from v_1 to v_2 (see Fig 9). The difference in velocity, Δv, can be found by subtracting v_1 from v_2. From the triangle in Fig 9, $\Delta v = v \times \Delta\theta$.

Acceleration is the rate of change of velocity so

$$a = \frac{\Delta v}{\Delta t} = \frac{v \times \Delta\theta}{\Delta t}$$

For a constant angular velocity,

$$\frac{\Delta\theta}{\Delta t} = \omega$$

This gives $a = v \times \omega$.

Since $v = r\omega$,

$$a = r\omega^2 = \frac{v^2}{r}$$

Examiners' Notes

The derivation of the expression for centripetal acceleration is given here for information. It will not be required in an exam.

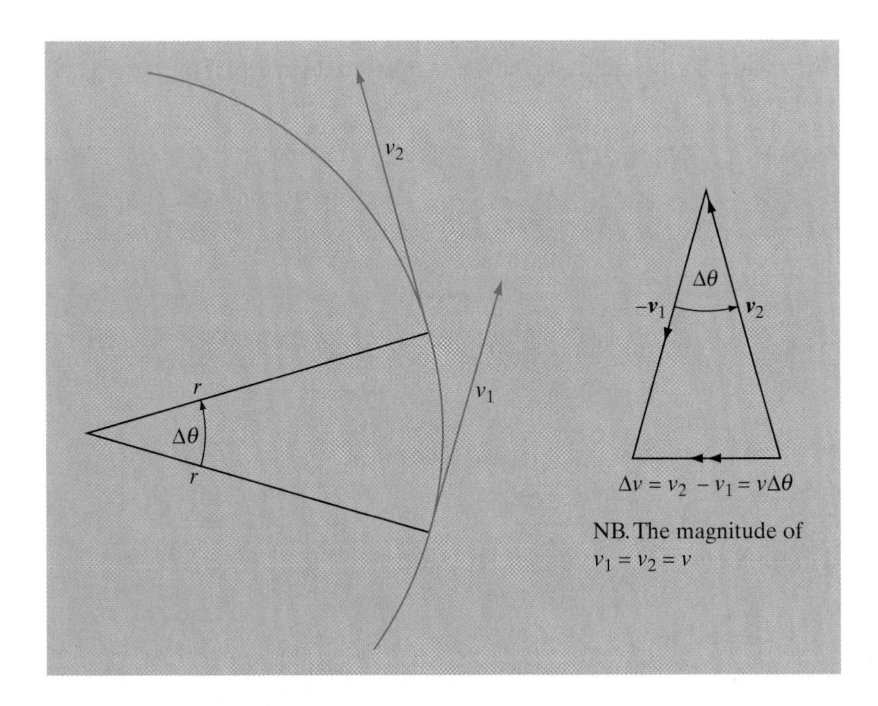

Fig 9
Deriving an expression for centripetal acceleration

Definition

The centripetal acceleration of an object moving in a circle is given by $a = r\omega^2 = v^2/r$.

Essential Notes

Note that the force is **centripetal** (towards the centre), NOT 'centrifugal'(away from the centre). People talk about centrifugal force 'throwing' them off their feet as a bus takes a sharp corner, or being responsible for spin-drying clothes. Using centrifugal force to explain these phenomena is a good way to make your teacher apoplectic. The person on the bus, or the water drops in a drier, are simply obeying Newton's First Law and continuing their motion in a straight line, until acted upon by a force.

Centripetal force

Since an object moving in a circle is accelerating there must be a resultant force acting on it. The resultant force could come from gravity, as in the case of a satellite orbiting the Earth, or from the tension in a string, as in the case of a conker being whirled round. The resultant force is **centripetal**; it acts towards the centre of the circle.

The size of the centripetal force that is needed to keep a mass, m, moving in a circle with a radius, r, at a velocity v, is given by Newton's Second Law in the form $F = ma$. The centripetal acceleration is v^2/r, so the centripetal force has to be

$$F = \frac{mv^2}{r} = mr\omega^2$$

Centripetal force:

- increases with mass, so that a larger force is needed to make a larger mass move at the same speed in a circle;
- increases with the square of the speed; this means that it takes four times as much friction to keep a car on the road if you take a corner twice as fast;
- decreases as the radius increases, so the force increases if the circle gets smaller.

Examiners' Notes

It is a common mistake in circular motion problems to invent an *extra* force called the centripetal force. Remember that this is not an extra force but simply the resultant of the *real* forces acting on the object in circular motion. Stick to real forces, with real causes.

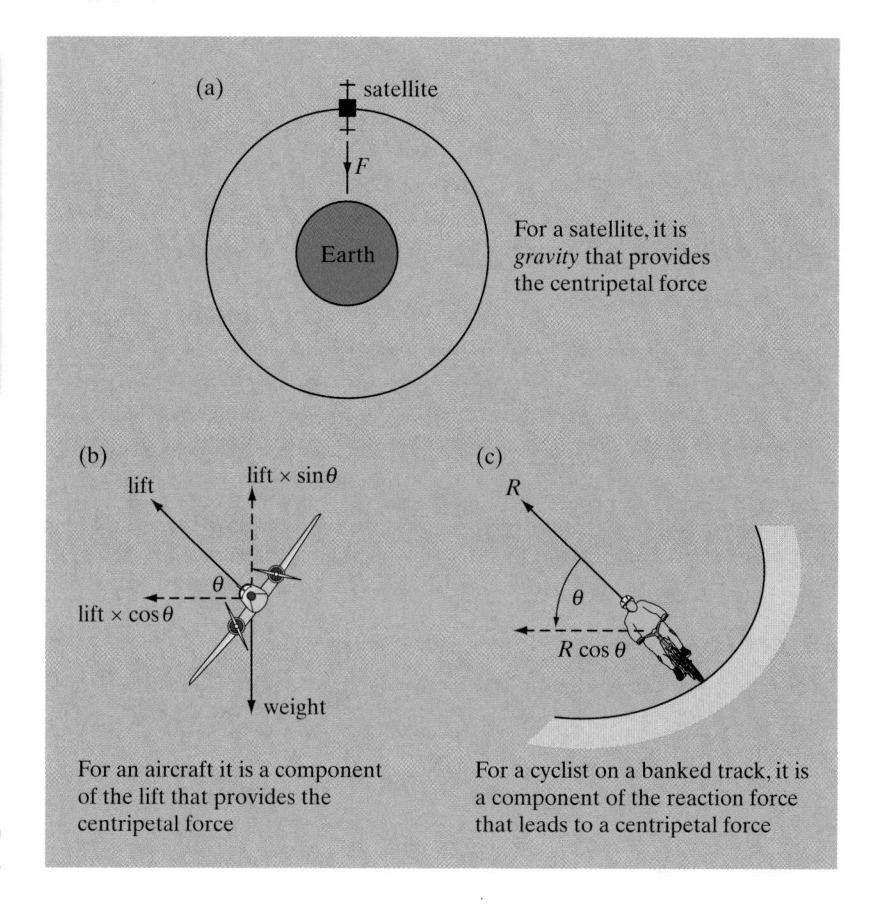

Fig 10
Examples of circular motion

Circular motion is unusual in that the force and the acceleration are not in the same direction as the velocity. In fact the force is always at right angles to the velocity. Because there is no motion in the direction of the force, there is no work done by the force. A satellite that is well above the Earth's atmosphere, like the Moon, can keep orbiting without any energy transfer taking place.

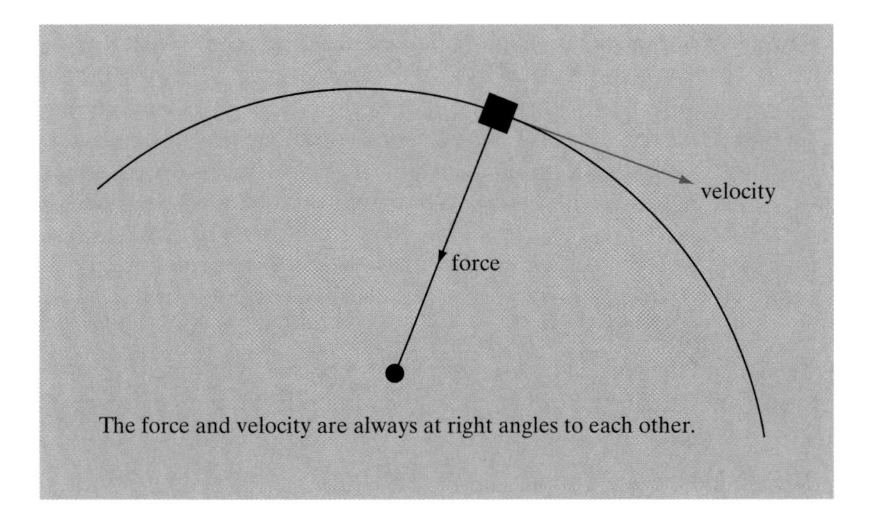

Fig 11
Force and velocity for circular motion

Example

A conical pendulum is a mass on a string that is whirled round in a horizontal circle.

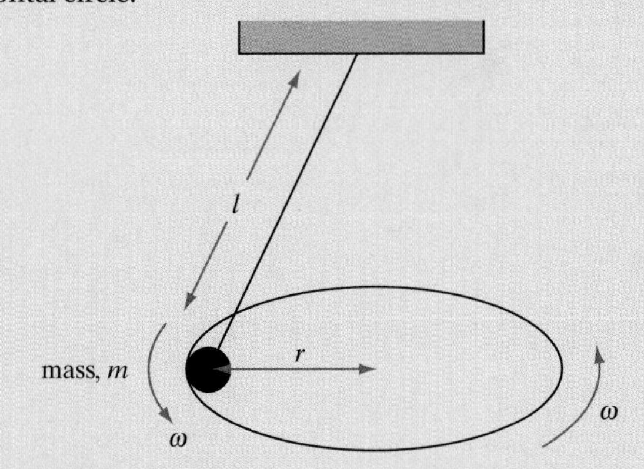

Fig 12

(a) Copy Fig 12 and mark in the forces that are acting on the mass.

(b) Identify the force or forces that are causing circular motion.

(c) If a 2 kg mass suspended on a 1 m long string is whirled around in a circle of radius 0.25 m, how fast will it be travelling?

(d) If the string broke, which direction would the mass move in at first?

Answer

(a) The forces that act on the mass are its weight, $W = mg$, and the tension, T, in the string.

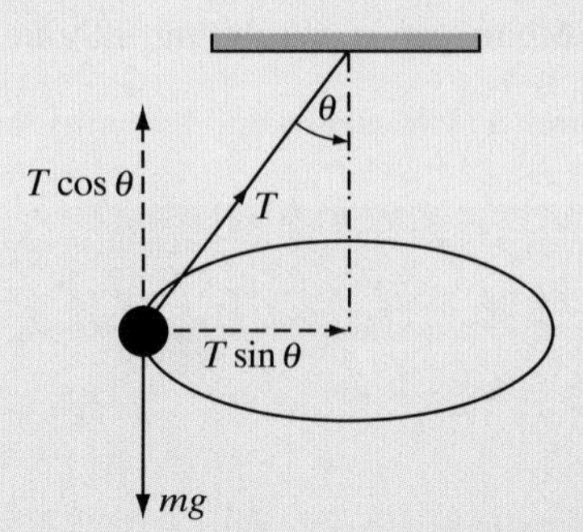

Fig 13

(b) Horizontally there is a resultant force of $T \sin \theta$ that acts towards the centre of the circle. This is the force that causes circular motion.

(c) The resultant force is $T \sin \theta = mv^2/r$.

We can find the tension by considering the vertical forces. Since the mass is moving in a horizontal circle, the vertical forces must balance: $T \cos \theta = mg$.

We can find the angle from

$$\sin \theta = \frac{r}{l} = \frac{0.25}{1} = 0.25, \quad \text{so} \quad \theta = \sin^{-1} 0.25 = 14.5°$$

Taking $g = 9.81\,\text{N}\,\text{kg}^{-1}$,

$$T = \frac{2\,\text{kg} \times 9.81\,\text{N}\,\text{kg}^{-1}}{\cos 14.5°} = 20.3\,\text{N}$$

This gives

$$v^2 = \frac{rT\sin\theta}{m} = \frac{0.25\,\text{m} \times 20.3\,\text{N} \times \sin 14.5°}{2\,\text{kg}} = 0.635\,\text{m}^2\,\text{s}^{-2}$$

The velocity is

$v = 0.797\,\text{m}\,\text{s}^{-1}$

(d) The mass would fly off at a tangent to the circle.

Simple harmonic motion

A car bouncing on its suspension, a child on a playground swing and the vibrations of a water molecule are all examples of **oscillations**. An oscillation is a repetitive, to-and-fro motion about a fixed position. This sort of motion is caused by a resultant force that is always directed to the same point. This resultant force, which changes direction as the object oscillates, is referred to as a **restoring force**.

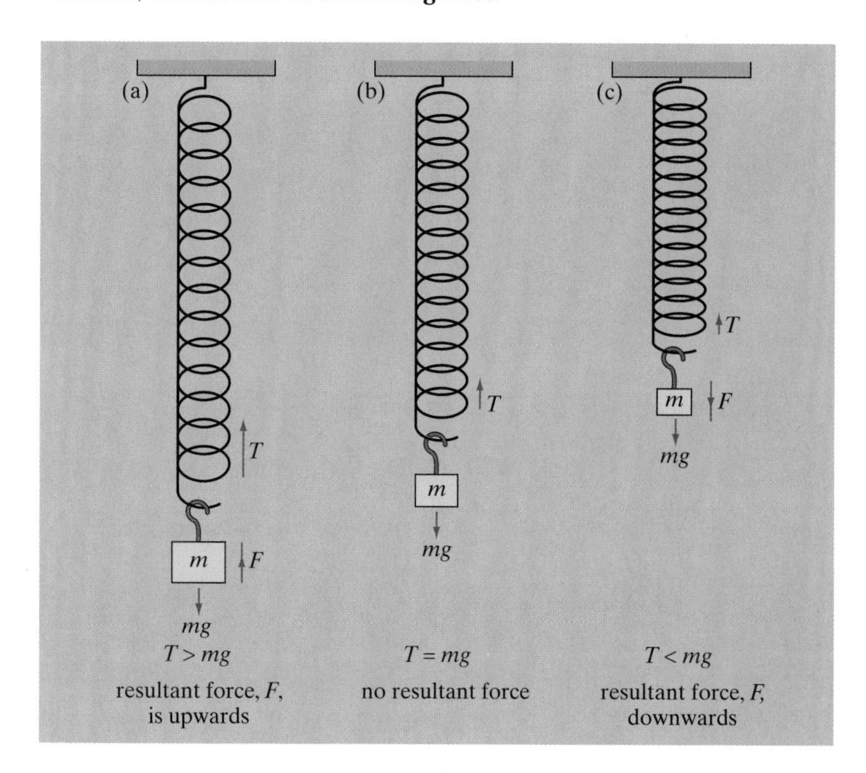

Fig 14
The restoring force on a mass suspended from a spring

Simple harmonic motion, or **SHM**, is a special case of oscillatory motion. If the magnitude of the restoring force on a body is proportional to its distance from the equilibrium position, we say that it moves with simple harmonic motion. Assuming that the mass of the object doesn't change, the acceleration will follow the same pattern.

Definition

The acceleration, a, of an object moving with simple harmonic motion, is always proportional to its displacement, x, from a fixed point. The acceleration is always directed towards that fixed point.

$$a \propto -x$$

Essential Notes

The negative sign in this relationship is there because the acceleration and the displacement are always in opposite directions, so that a positive displacement causes a negative acceleration, and vice versa.

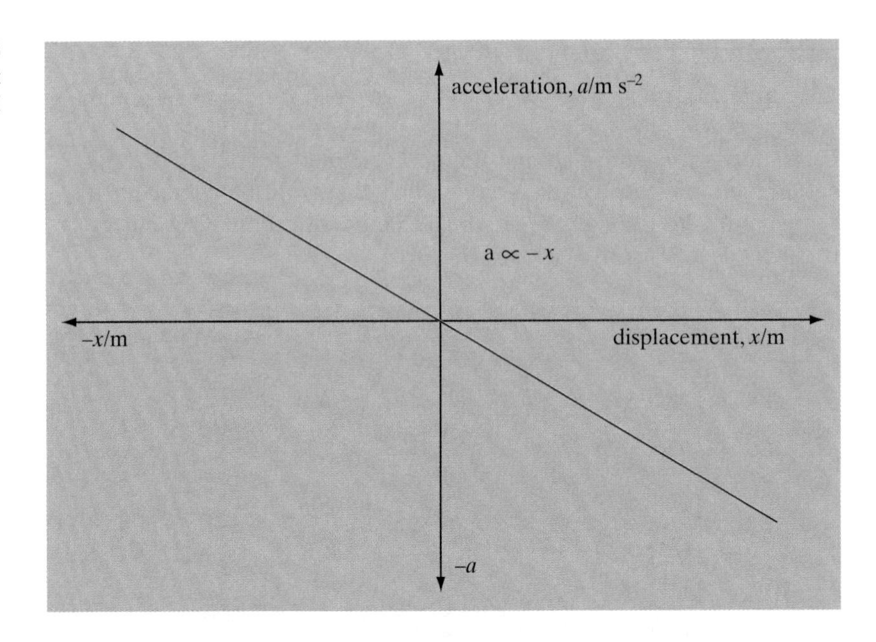

Fig 15
Acceleration against displacement for an object moving with SHM

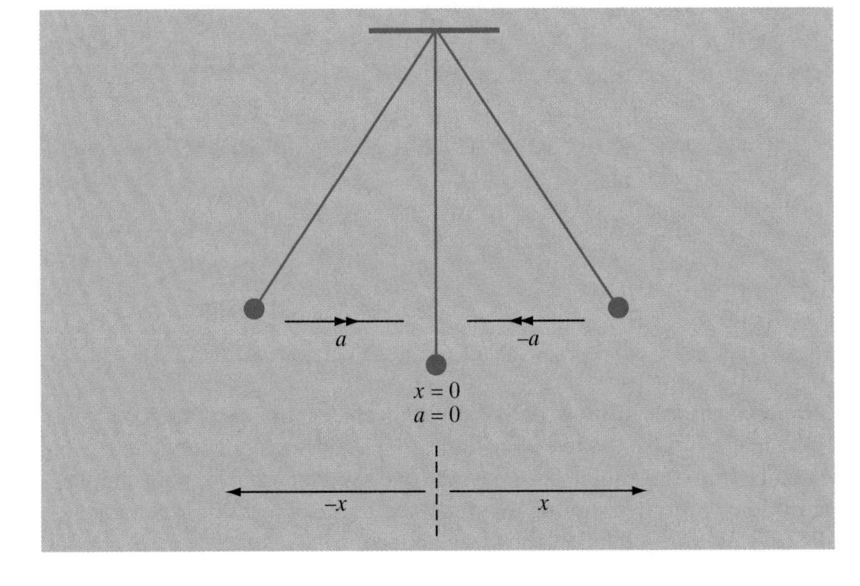

Fig 16
Acceleration (a) and displacement (x) for a pendulum

Essential Notes

It can be difficult to realise that the pendulum has its greatest acceleration when it isn't moving at all. Conversely, the pendulum has no acceleration when it is moving fastest.

A swinging pendulum is an example of a body moving with SHM. The restoring force, and the acceleration, is greatest when the pendulum is furthest from equilibrium, when the **displacement** is equal to its maximum value or **amplitude**, A. The acceleration drops to zero when the displacement of the pendulum is zero.

Definitions

***Displacement** is the distance from a fixed point in a certain direction. It is a vector quantity.*

***Amplitude** is the magnitude of the maximum displacement. It is a scalar quantity.*

Period, frequency and acceleration

Oscillations are often referred to as **periodic motion**. The vibrations of a guitar string or the motion of a piston in a car engine are examples of periodic motion. In each case a pattern of motion is repeated over and over again. The time taken to complete one full cycle of motion is called the **period**, T.

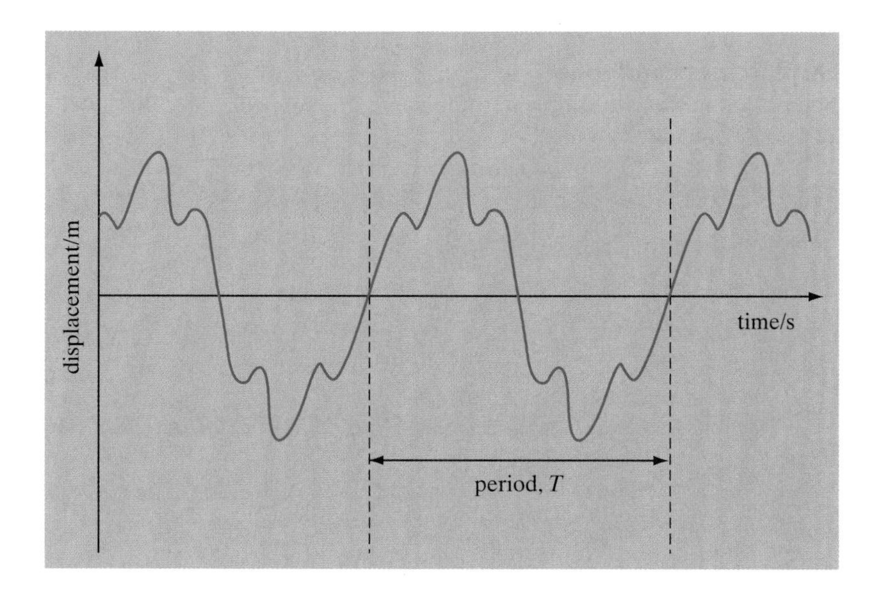

Fig 17
Periodic motion

The number of oscillations that an object completes in one second is called the **frequency** of the oscillation. Frequency is measured in hertz, Hz. Frequency is related to period by the expression:

$$\text{frequency} = \frac{1}{\text{period}} \qquad f = \frac{1}{T}$$

High-frequency oscillations tend to involve high accelerations, since the body has to speed up and slow down many times per second. For simple harmonic oscillations the acceleration is linked to the frequency and the displacement by the equation

$$a = -(2\pi f)^2 x$$

Example

The piston in a car engine moves with a motion that is approximately simple harmonic. One cycle of motion takes 0.017 s and the piston moves through a total distance of 100 mm. Calculate the maximum acceleration of the piston.

Answer

The piston's frequency is: $f = \dfrac{1}{0.017\,\text{s}} = 58.8\,\text{Hz}$

The maximum acceleration will be when the piston reaches its maximum displacement. This will be equal to 50 mm (half the total movement of the piston), so

$$a = -(2\pi f)^2 x = -(2\pi \times 58.8)^2 \times 0.05 = 6800\,\text{m s}^{-2} \text{ (to 2 s.f.)}$$

in a direction opposite to that of the displacement.

Displacement and time

Bodies that move with simple harmonic motion have a displacement that varies sinusoidally with time.

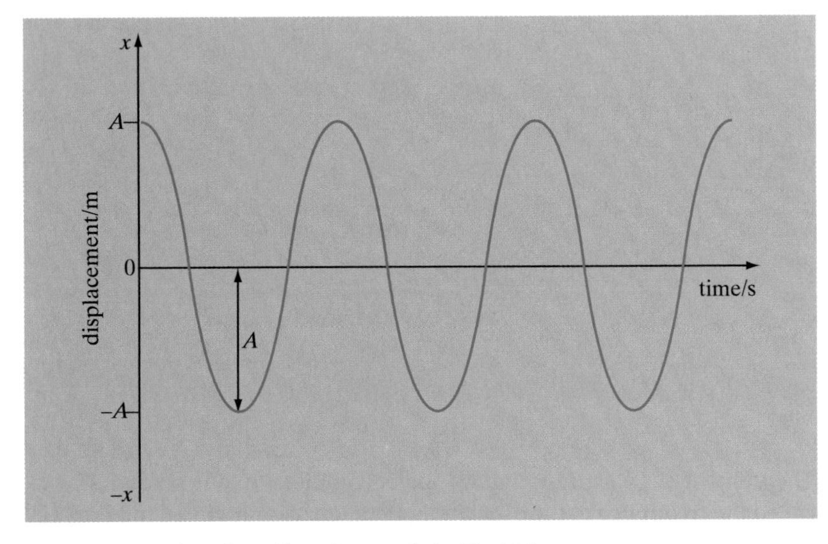

Fig 18
Displacement against time for a simple harmonic oscillator

Essential Notes

To convert degrees to radians you need to remember that there are 2π radians in a circle. So 2π radians = 360°. From this you can work out that 1 radian = $360/2\pi$ = 57.3°. See Table 1, page 10.

The equation that describes the graph in Fig 18 is

$$x = A\cos(2\pi ft)$$

where x is the displacement, A is the amplitude, f is the frequency and t is the time. The term $2\pi ft$ has units of radians (see page 9).

Example

(a) Plot a displacement–time graph to show three cycles of an oscillation that has an amplitude of 6 mm and a frequency of 5000 Hz.

(b) Calculate the displacement after 0.24 m s.

Answer

(a) You need to use the equation

$$x = A\cos(2\pi ft)$$

$$x = 0.006\cos(10000\pi t)$$

Now you need to choose suitable values of t to calculate values of x for plotting. Values of t = 1, 2, 3, etc. are too large and represent points which are thousands of oscillations apart.

The frequency of the oscillation is 5000 Hz, so the oscillation has a period of $1/f = 1/5000 = 0.0002$ s or 0.2 ms.

The oscillation will have a maximum displacement at 0, 0.2, 0.4, 0.6 ms etc.

The oscillation will have a minimum (negative) displacement at 0.1, 0.3, 0.5 ms etc.

The oscillation will have zero displacement in between these points, at 0.05, 0.15, 0.25, 0.35 ms, etc.

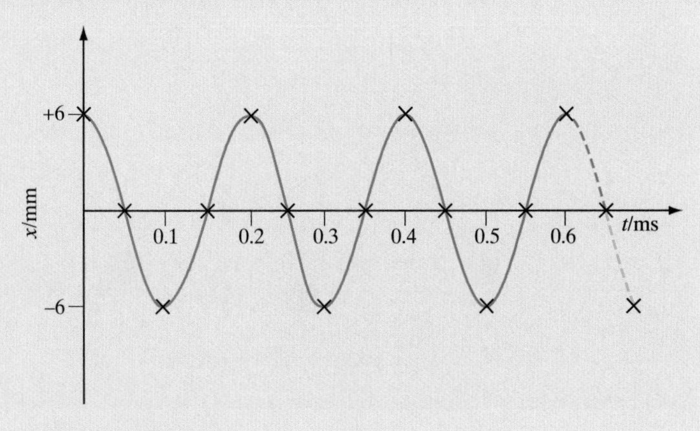

Fig 19

(b) $x = 0.006 \cos (10\,000\, \pi t)$

$x = 0.006 \cos (10\,000 \times \pi \times 0.24 \times 10^{-3})$

$x = 1.9 \times 10^{-3}$ m $= 1.9$ mm

Examiners' Notes

Remember to set your calculator to radians when you calculate this formula.

Velocity and time

The velocity of a particle that is moving with SHM can be found from the gradient of the displacement vs time graph.

Essential Notes

Remember (from Unit 2) that velocity $= \Delta x/\Delta t$.

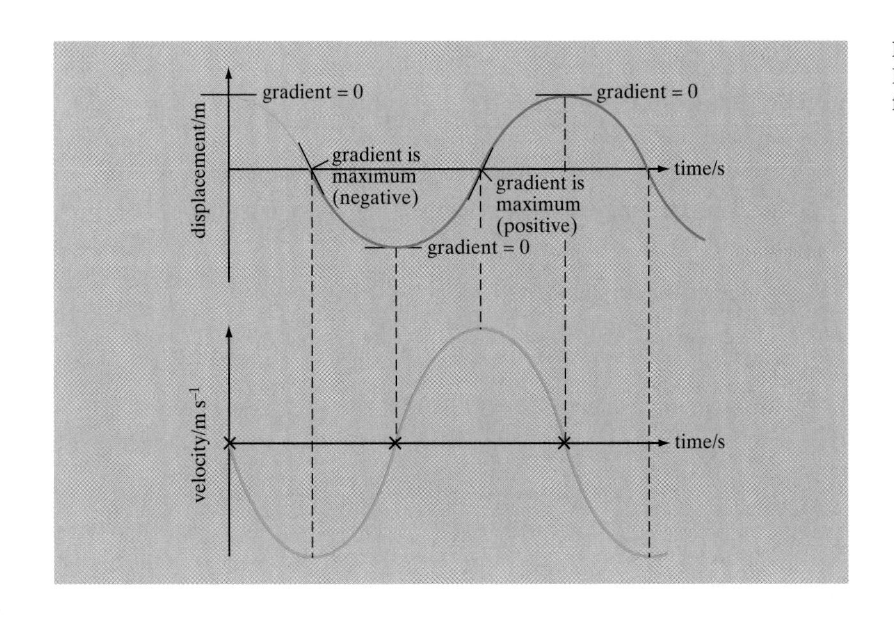

Fig 20
Displacement and velocity vs time for a simple harmonic oscillator

The velocity of a simple harmonic oscillator reaches a maximum when the displacement is zero. The velocity is zero when the displacement is at a maximum. Think of a pendulum where the velocity is zero for an instant at each end of the pendulum's swing, and the velocity of the pendulum bob is at its greatest when it swings through the equilibrium position.

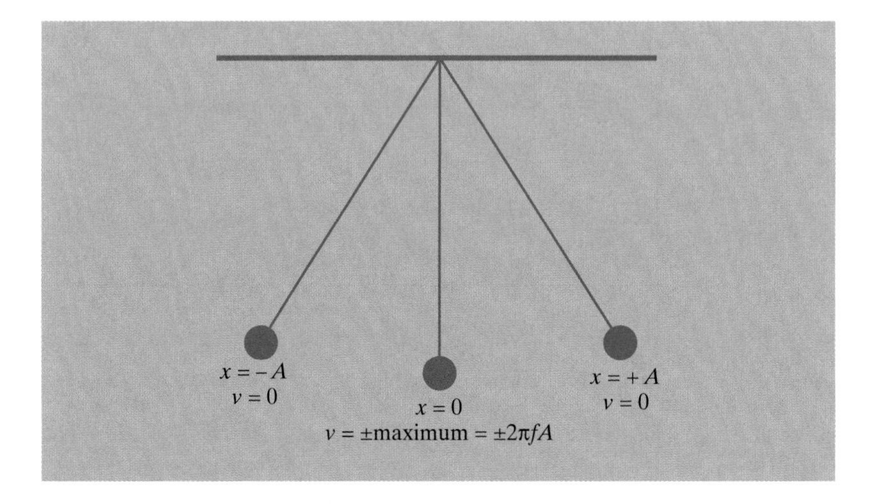

Fig 21
Velocity and displacement for a pendulum bob

This pattern of changing velocity can be linked to the displacement by the equation

$$v = \pm 2\pi f\sqrt{A^2 - x^2}$$

Example

A mass suspended from a spring oscillates with a period of 1.5 s. The amplitude of the oscillations is 5 cm. Find

(a) the maximum velocity

(b) the minimum velocity

(c) the velocity when the displacement is half the amplitude

(d) the displacement when the velocity is half of its maximum value.

Answer

The period of the oscillation is 1.5 s, so the frequency is

$$f = \frac{1}{1.5} = 0.67\,\text{Hz}$$

(a) The maximum velocity occurs when there is no displacement, $x = 0$:

$$v = \pm 2\pi f\sqrt{A^2 - x^2} = \pm 2\pi f A$$
$$= \pm 2\pi \times 0.67 \times 0.05 = 0.21\,\text{m}\,\text{s}^{-1}$$

(b) The minimum velocity is when $x = A$, and then $v = 0\,\mathrm{m\,s^{-1}}$.

(c) The velocity at $x = 0.025\,\mathrm{m}$ is

$$v = \pm 2\pi \times 0.67 \times \sqrt{0.05^2 - 0.025^2} = \pm 0.18\,\mathrm{m\,s^{-1}}$$

(d) When the velocity is $\frac{1}{2}$ of its maximum value, $v = 0.105\,\mathrm{m\,s^{-1}}$.

$$v^2 = 4\pi^2 f^2(A^2 - x^2) \quad \text{so} \quad x^2 = A^2 - \frac{v^2}{4\pi^2 f^2}$$

This gives $x = \pm 0.043\,\mathrm{m}$ from equilibrium.

Essential Notes

Notice that the velocity does not reach ½ of its maximum value at the midpoint between zero and maximum displacement.

Acceleration

Acceleration is the rate of change of velocity, so an oscillating object will have its greatest acceleration when the velocity is changing most quickly, at the ends of the oscillation.

Essential Notes

Remember that acceleration = $\Delta v/\Delta t$.

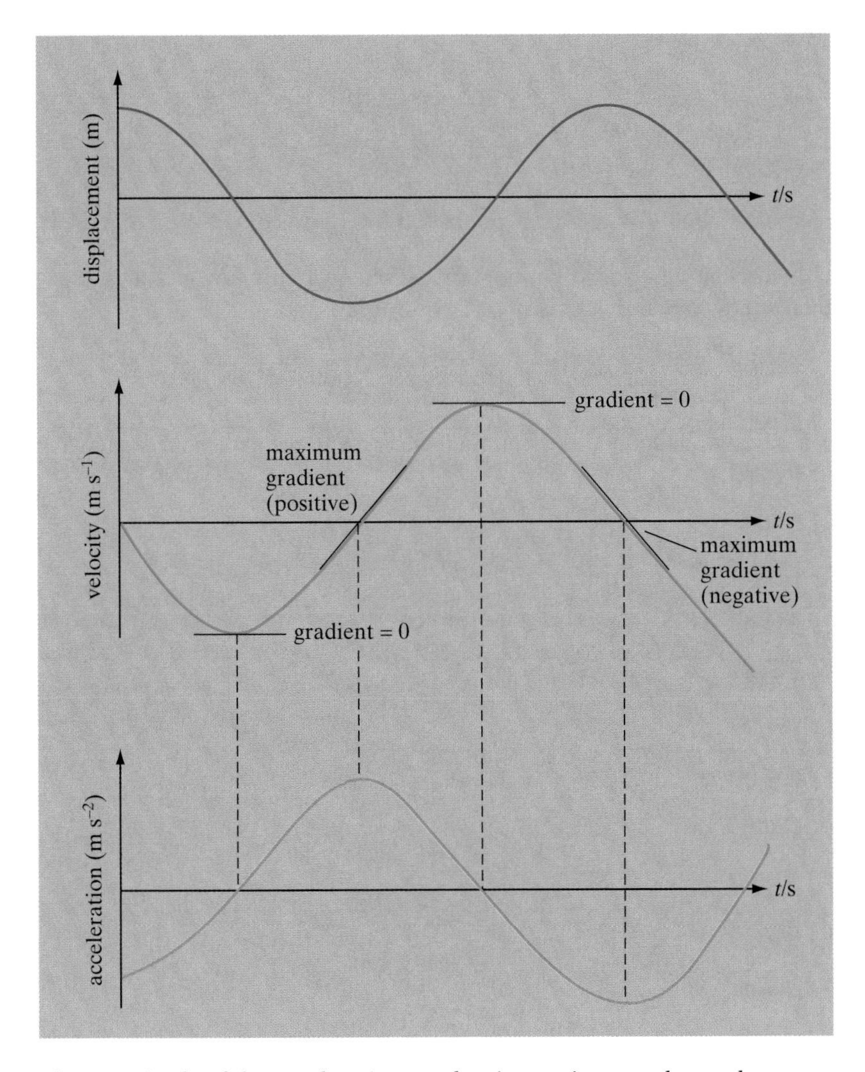

Fig 22
Displacement, velocity and acceleration vs time graph for a simple harmonic oscillator

The magnitude of the acceleration reaches its maximum value at the same time as the displacement, though they are in opposite directions.

Energy in oscillations

Objects that oscillate are continually transferring energy from potential energy to kinetic energy and back again. A pendulum uses its kinetic energy to do work against gravity, storing gravitational potential energy. This energy is then transferred as kinetic energy as the pendulum accelerates back towards the centre of the oscillation.

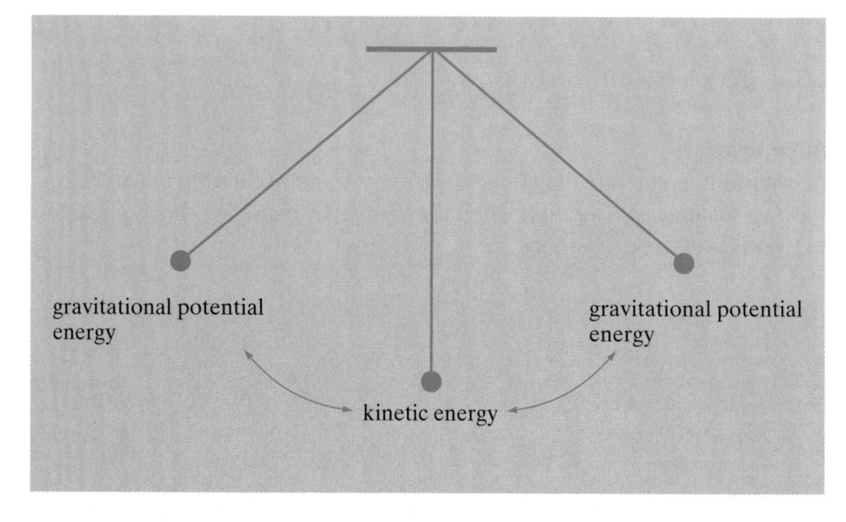

Fig 23
Energy transfer during the oscillation of a pendulum

If the oscillating system does not transfer energy to the surroundings, the total energy remains constant:

Total energy = kinetic energy + potential energy

The variation of kinetic and potential energy with displacement is shown in Fig 24.

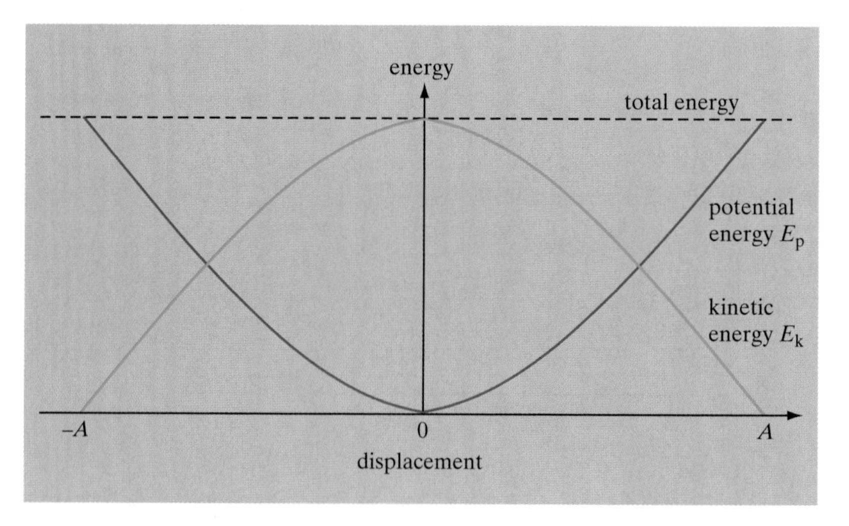

Fig 24
Energy vs displacement for a pendulum oscillation

To see how the energy of a mass moving with SHM changes with *time*, imagine a mass and a spring placed on a frictionless, horizontal surface (Fig 25). The mass is pulled to the right, to a maximum value of displacement, A, and then released. The mass oscillates in simple harmonic motion about the equilibrium position, compressing and extending the spring. At each end of the oscillation, when $x = \pm A$, the system has potential energy, stored as elastic strain energy in the spring. As the mass passes through the equilibrium position, when the spring is at its natural length, all the energy is kinetic.

Fig 25
A horizontal mass–spring system

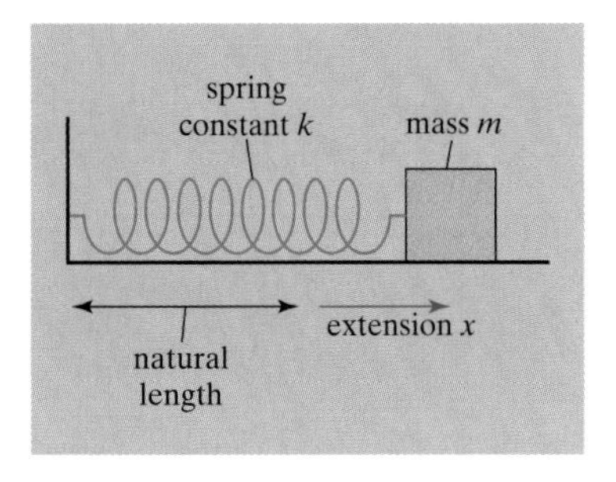

The kinetic energy of the mass is given by

$$E_k = \tfrac{1}{2} mv^2$$

Since this is SHM, $v^2 = 4\pi^2 f^2(A^2 - x^2)$, so

$$E_k = 2m\pi^2 f^2(A^2 - x^2) = 2m\pi^2 f^2 (A^2 - A^2\cos^2 2\pi ft) = 2m\pi^2 f^2 A^2 (1 - \cos^2 2\pi ft)$$

Since $1 - \cos^2\theta = \sin^2\theta$, the kinetic energy is given by:

$$E_k = 2m\pi^2 f^2 A^2 \sin^2 2\pi ft$$

So for a given mass, oscillating at a steady frequency f, the kinetic energy is proportional to $\sin^2 2\pi ft$.

The potential energy is stored in the spring as elastic strain energy:

$$E_p = \tfrac{1}{2} kx^2$$

where k is the spring constant. For SHM, $x = A \cos 2\pi ft$, so

$$E_p = \tfrac{1}{2} kA^2 \cos^2 2\pi ft$$

The potential energy is therefore proportional to $\cos^2 2\pi ft$.

Examiners' Notes

Remember from Unit 2 that the energy stored in a spring is $\frac{1}{2}Fx$, where x is the extension, and since Hooke's Law gives $F = kx$, the energy stored is equal to $\frac{1}{2}kx^2$.

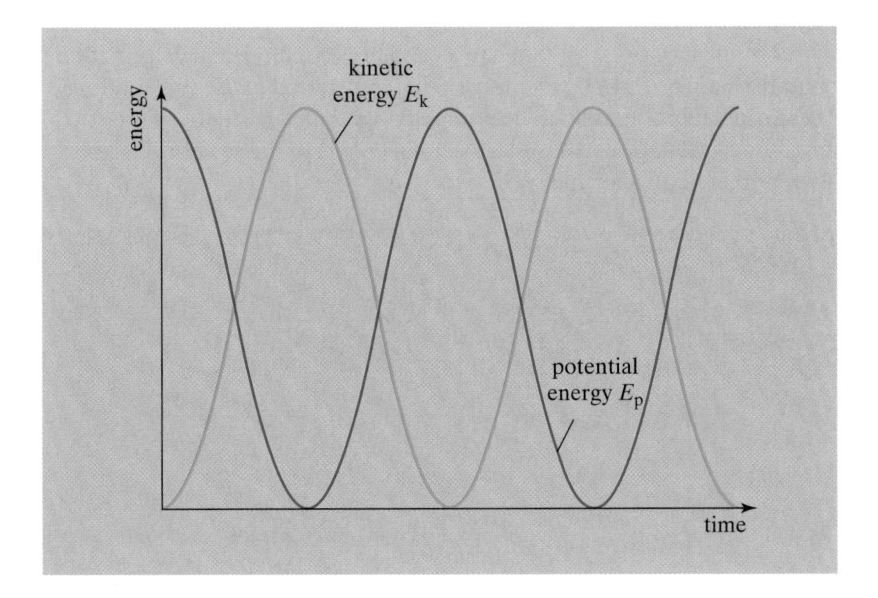

Fig 26
The variation of kinetic energy and potential energy with time over one cycle of SHM

Example

A mass of 500 g is hung from a spring which extends by 6 cm. The mass is pulled down a further 4 cm and then released. Calculate the maximum kinetic energy of the mass, and hence find its maximum velocity.

Answer

The energy stored in a stretched spring is $\frac{1}{2}kx^2$, where k is the spring constant.

$$k = \frac{\text{force}}{\text{extension}} = \frac{4.91\,\text{N}}{0.05} = 81.8\,\text{N m}^{-1}\ (\text{taking } g \text{ as } 9.81\,\text{N kg}^{-1})$$

The potential energy stored in the spring when it is pulled down by a further 4 cm, when the total extension is 10 cm, is

$$E = \tfrac{1}{2} \times 81.8 \times 0.10^2 = 0.409\,\text{J}$$

This is transferred as kinetic energy and gravitational potential energy when the spring is released. When the mass passes back through the equilibrium position, the potential energy gained is

$$\Delta E_p = mg\,\Delta h = 0.5 \times 9.81 \times 0.04 = 0.196\,\text{J}$$

The energy still stored in the spring is $E = \frac{1}{2} \times 81.8 \times 0.06^2 = 0.147\,\text{J}$

Therefore the kinetic energy is the remainder:

$$E_k = 0.409\,\text{J} - 0.196\,\text{J} - 0.147\,\text{J} = 0.066\,\text{J}$$

Because kinetic energy is $E_k = \frac{1}{2}mv^2$, the velocity is

$$v = \sqrt{\frac{2E_k}{m}} = 0.51\,\text{m s}^{-1}$$

Damped oscillations

In real oscillating systems there are always resistive forces, such as friction or air resistance, that lead to energy being transferred to the surroundings. These oscillations are said to be **damped**. In damped oscillations the amplitude gradually decreases as the total energy of the system gets less. The period remains unchanged.

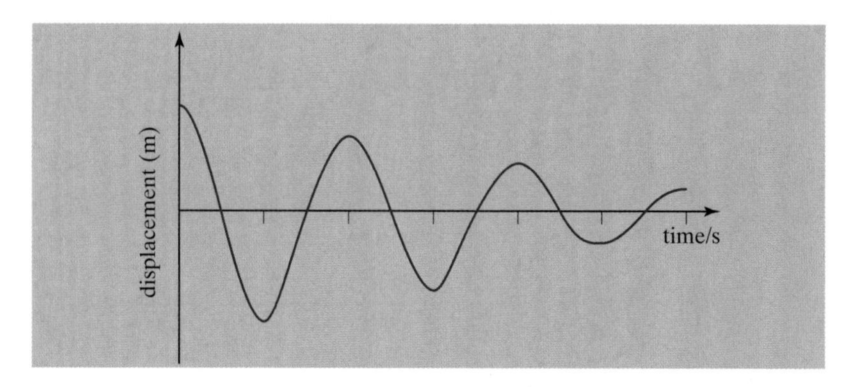

Fig 27
Damped oscillations

Simple harmonic systems

The simple pendulum

A 'simple' pendulum is one where all the mass is concentrated at one end. This mass is called the pendulum 'bob'. The mass of a simple pendulum's suspension is small compared to the bob's mass and so we can neglect it. When this is not true, the pendulum is referred to as a compound pendulum.

We can show that an oscillating simple pendulum is an example of simple harmonic motion. To do this we need to prove that the acceleration is proportional to the displacement from equilibrium. First we need to consider the forces acting on the pendulum. There are only two forces acting, the weight of the bob, W, and the tension in the string, T. We resolve the weight into two components, acting parallel and perpendicular to the tension (Fig 29).

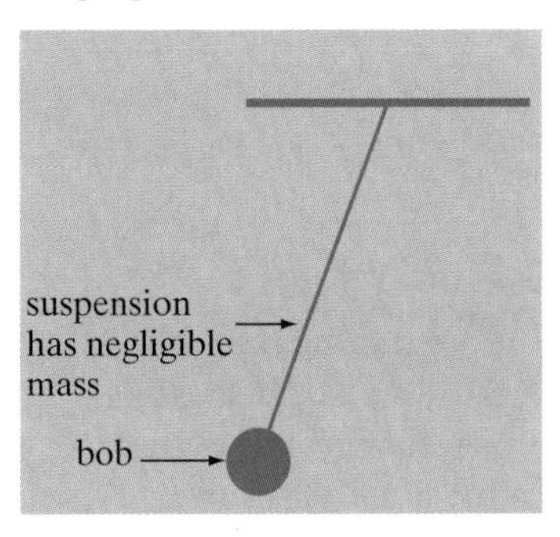

Fig 28
Simple pendulum

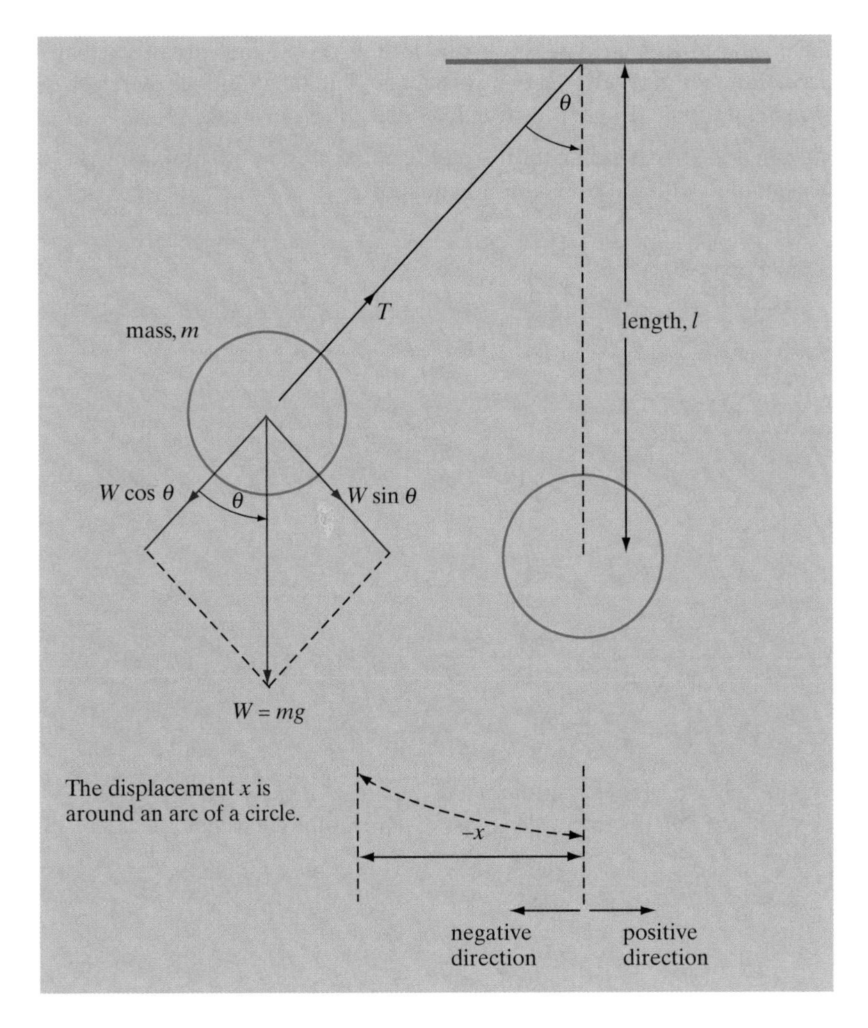

Fig 29
Forces acting on the bob of a simple pendulum

Essential Notes

If the angle θ is small enough, we can assume that θ in radians is approximately equal to $\sin\theta$. (This is valid up to about 10°, $\sin 10° = 0.17364$ and θ (radians) $= 0.17453$, a difference of only about 0.5%.)

Perpendicular to the string, there is only one force acting. This will provide an acceleration back towards the equilibrium position.

The resultant force is $W \sin \theta = mg \sin \theta$.

We can use Newton's Second Law to find the acceleration of the bob:

$$mg \sin\theta = ma$$

or

$$a = g \sin\theta$$

Essential Notes

The negative sign appears because we have defined displacement to the left as negative, and a negative displacement produces an acceleration to the right.

If the angle θ is small, we can say that $\sin \theta \cong \theta = -x/l$.

So

$$a = -\left(\frac{g}{l}\right)x$$

The definition of SHM states that the acceleration must be proportional to the displacement. Since g/l is constant for this pendulum, we can say that acceleration is proportional to displacement and the motion is therefore simple harmonic. This is only true for small angles of swing.

We can find an expression for the period of the motion by comparing the acceleration with the general SHM equation.

Simple pendulum $\quad a = -\left(\frac{g}{l}\right)x$

General SHM equation $\quad a = -(2\pi f)^2 x$

We can see that $\quad \frac{g}{l} = (2\pi f)^2$

So $\quad f = \frac{1}{2\pi}\sqrt{\frac{g}{l}}$

As $T = 1/f$,

$$T = 2\pi\sqrt{\frac{l}{g}}$$

The period of a simple pendulum just depends on its length and on the acceleration due to gravity. The mass of a pendulum does not affect its period.

Example

A child on a playground swing oscillates with a motion that is approximately simple harmonic.

(a) If the child has a mass of 30 kg and the swing is 2 m long, calculate the period of oscillation.

(b) The child's younger sister, mass 20 kg, now gets onto the swing instead. What difference would you expect to the period of the pendulum?

Answer

(a) We use the equation $T = 2\pi\sqrt{\frac{l}{g}}$

$$T = 2\pi\sqrt{\frac{2}{9.81}} = 2.8\,\text{s (to 2 s.f.)}$$

(b) The period will be almost unchanged. The period of a pendulum does not depend on its mass. The effective length of the pendulum would be altered if its centre of mass was different, and this would affect the period slightly.

Mass on a spring

A mass bouncing on a spring is another example of simple harmonic motion.

To show that this is SHM we need to consider the forces acting on the mass.

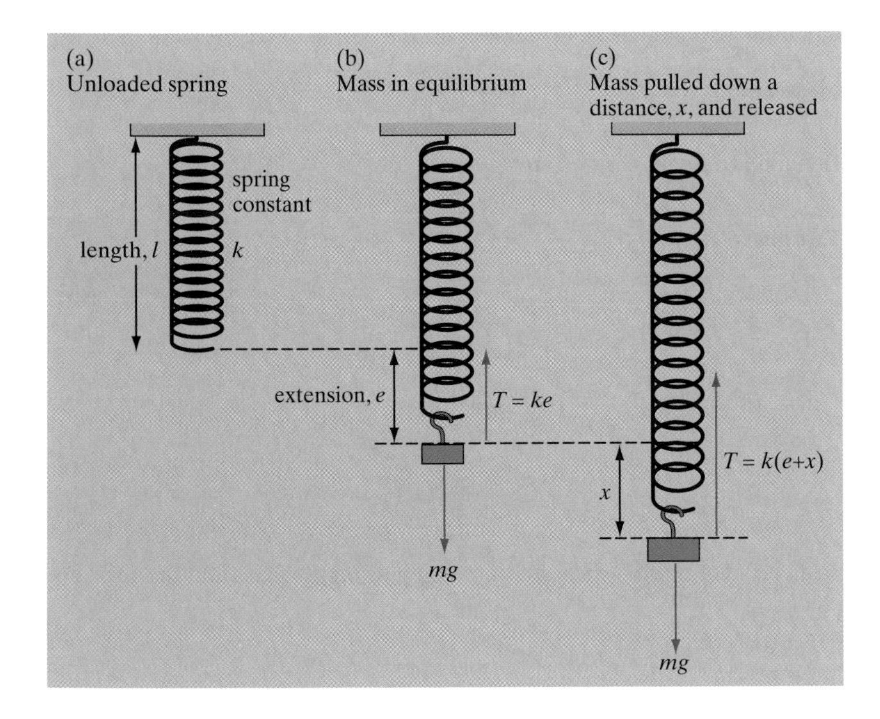

Fig 30
The forces acting on a mass on a spring

When the mass is in equilibrium the restoring force from the spring, T, balances the weight, W:

$$W = T \quad \text{or} \quad mg = ke$$

When the mass is pulled down by a further displacement x, there is a net force upwards of

$$k\,(e + x) - mg$$

This causes an upwards acceleration of a.

Using Newton's Second Law:

$$-(k\,(e + x) - mg) = ma$$

Since $mg = ke$,

$$-\,kx = ma$$

this gives

$$a = -\left(\frac{k}{m}\right)x$$

Essential Notes

The negative sign shows that the net force is upwards, whereas the displacement x is measured from the equilibrium position in the downwards direction.

For any given mass and spring, m and k are constants, so the acceleration is proportional to the displacement and directed in the opposite direction. The motion is therefore simple harmonic.

If we compare the equation for acceleration with the general SHM equation, we can find an expression for the period of the oscillations.

Mass–spring system $\quad a = -\left(\frac{k}{m}\right)x$

General SHM equation $\quad a = -(2\pi f)^2 x$

This gives $\quad \frac{k}{m} = (2\pi f)^2$

So $\quad f = \frac{1}{2\pi}\sqrt{\frac{k}{m}}$

As $T = 1/f$,

$$T = 2\pi\sqrt{\frac{m}{k}}$$

The period of a mass–spring system depends on the mass and the spring constant.

Example

A family car has a mass of 1000 kg when it is not loaded. This mass is supported equally by four springs. When the car is fully loaded its mass goes up to 1250 kg and the springs compress by a further 2 cm. When the car goes over a bump in the road, it bounces on its springs. Find the period of these oscillations.

Answer

The formula for the period of a mass–spring system is $T = 2\pi\sqrt{\frac{m}{k}}$.

We know the mass of the system but we need to calculate the spring constant, k.

The extra weight of $250\,\text{kg} \times 9.81\,\text{N kg}^{-1} = 2450\,\text{N}$, will depress the four springs by 0.02 m.

If each spring carries $\frac{1}{4}$ of the weight, we can use Hooke's Law, $F = ke$, to find the spring constant of one spring:

$$k = \frac{F}{e} = \frac{613}{0.02} = 3.06 \times 10^4\,\text{N m}^{-1}$$

Then we can find the period of oscillation of a spring. The effective mass oscillating on each spring is $1250/4 = 313$ kg.

$$T = 2\pi\sqrt{\frac{m}{k}} = 2\pi\sqrt{\frac{313}{3.06 \times 10^4}} = 0.64\,\text{s}$$

Forced vibrations and resonance

When a child's swing is pulled back and released, it will oscillate at a frequency that just depends on the strength of gravity and the length of the swing. The swing is said to oscillate at its **natural frequency**, f_0. These free vibrations are not influenced by any externally varying force.

Definition

The natural frequency of an oscillating system is the frequency at which the system undergoes free vibrations.

When a loudspeaker vibrates in response to the oscillating electrical signal that drives it, it is undergoing **forced vibrations**. These forced vibrations happen at the **driving frequency** of the electrical signals. If the driving frequency matches the natural frequency of the loudspeaker, large-amplitude vibrations will occur, which could lead to an unpleasant noise coming from the loudspeaker. This phenomenon is known as **resonance**.

Definition

Resonance occurs when a system is forced to oscillate at its natural frequency, i.e. when the driving frequency equals the natural frequency of the system.

Resonance often causes problems in mechanical systems. In 1942 the Tacoma Narrows Bridge in the USA collapsed spectacularly when turbulence from the wind set it vibrating. The frequency of the forces from the wind eddying around the bridge matched the natural frequency of the suspension bridge and large-amplitude oscillations built up, eventually destroying the bridge. Another example of a bridge collapsing through resonance happened in Angers, France in 1850. This time the driving force came from soldiers marching across the bridge. The frequency of the marching matched a natural frequency of the bridge, resonance occurred, the bridge collapsed and 200 soldiers died.

Resonance can also be useful. In microwave cooking it is the resonant vibrations of water molecules that heat the food. The driving frequency of the electromagnetic waves matches a natural vibration frequency of water molecules and large-amplitude vibrations occur.

Effects of damping

A familiar example of resonance occurs in a children's playground when an adult is pushing a child on a swing. The adult's pushing frequency matches the natural frequency of the swing and the amplitude of the oscillations gets larger. With each push the adult is transferring energy to the swing. If this happened over many cycles the amplitude could get dangerously large. It is **damping** which prevents the amplitude of the oscillations continually increasing. Damping transfers energy from the oscillating system to the surroundings. In the case of a swing, the resistive forces of friction at the support and air resistance acting on the child will limit the size of the oscillations. In theory, if there was no damping, a resonant system would increase its energy every cycle and the amplitude would also keep increasing.

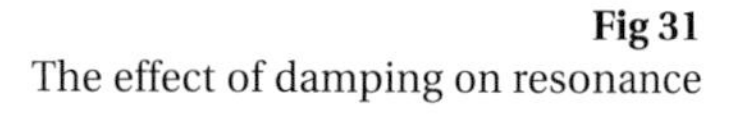

Fig 31
The effect of damping on resonance

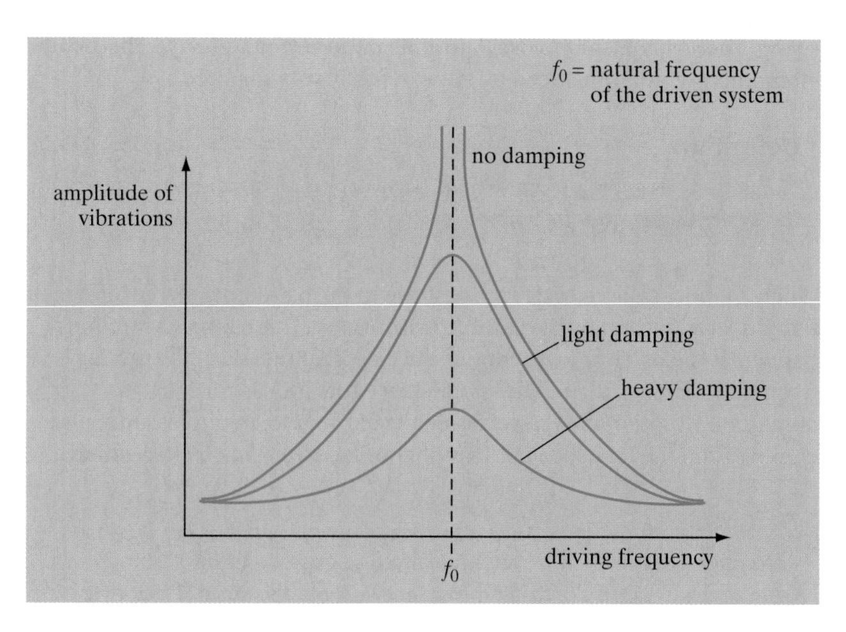

Example

The casing of a washing machine vibrates as the motor driving the drum spins round. As the motor speeds up the vibrations increase until, at a certain motor speed, the casing vibrates violently. At higher motor speeds the vibrations die away again.

(a) Explain these observations.

(b) Washing machines often have a large block of concrete bolted to the casing. Explain why.

(c) How could the amplitude of these vibrations be decreased?

Answer

(a) The casing is resonating. When the driving frequency of the oscillations caused by the motor matches the natural vibration frequency of the casing, there are large-amplitude vibrations. At lower and higher motor speeds there is less energy transferred to the casing during each cycle and the vibrations get less.

(b) The concrete increases the mass of the vibrating system and lowers its natural frequency so that resonance does not occur at the motor's normal running speed.

(c) Adding extra damping to the system can decrease the amplitude. A shock absorber can be fitted to the casing. The shock absorber is a piston moving in a cylinder filled with oil. As the casing vibrates, oil is forced through small holes in the piston, dissipating energy and reducing the vibrations.

Phase difference of driven systems

Imagine that you are holding a spring attached to a mass, and that you move your hand up and down (Fig 32). At very low frequencies the mass moves in step with your hand: there is no **phase difference** between the driver and the driven system. If you vibrate your hand faster, the mass will lag behind, and as you approach the natural frequency of the mass–spring system there is a 90° phase difference between the driving force and the driven oscillations.

This phase lag can be demonstrated by the apparatus known as Barton's pendulums (see Fig 33). A heavy pendulum, the driver, is attached to a string which also supports a number of other, lighter pendulums of different lengths. When the driver is set in motion, some of the energy is transferred to the other pendulums. The light pendulum that has the same length as the driver oscillates with a large amplitude, that is it resonates, since it has the same natural frequency.

The resonating pendulum is 90°, or $\pi/2$ radians, behind the driving pendulum. The shorter pendulums are almost in phase with the driver; the longer ones are almost 180° or π radians out of phase with the driver.

Fig 32
Driving oscillations on a spring

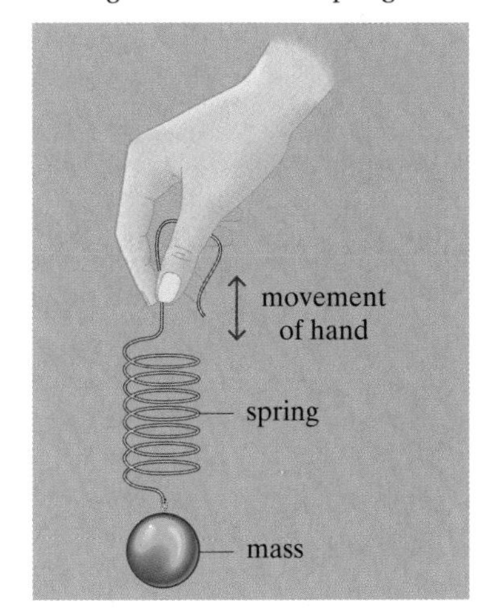

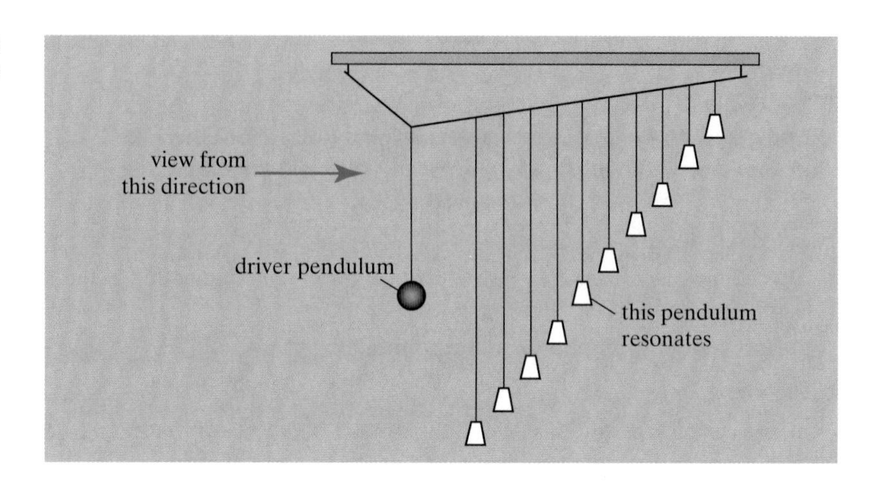

Fig 33
Barton's pendulums

Essential Notes

When two continuous similar waves are travelling in opposite directions, they can superpose to form a **stationary wave** (see Unit 2). A stationary wave is a fixed pattern of vibration. Unlike a progressive wave, no energy is transferred along the wave.

Stationary waves

Stationary waves are an example of resonance. Those on a string fixed at each end, for example, occur when the string oscillates with an integral number of half-wavelengths along its length.

Stationary waves on the strings of an instrument, such as a guitar or a violin, are the source of vibration for musical notes. When the string is made to vibrate, by being plucked or scraped with a bow, reflections from either end superpose to cause the stationary wave. Since a string is fixed at both ends, there must be a vibration node at each end. The simplest way that a string can vibrate is with one antinode in the middle of the string. This wave pattern is known as the **fundamental mode**. The frequency of the fundamental mode, f_0, is given by c/λ. If the string has a length l, the wavelength λ is $2l$, since there is half a wave on the string. This gives the fundamental frequency as

$$f_0 = \frac{c}{2l}$$

where c is the speed of the wave along the string.

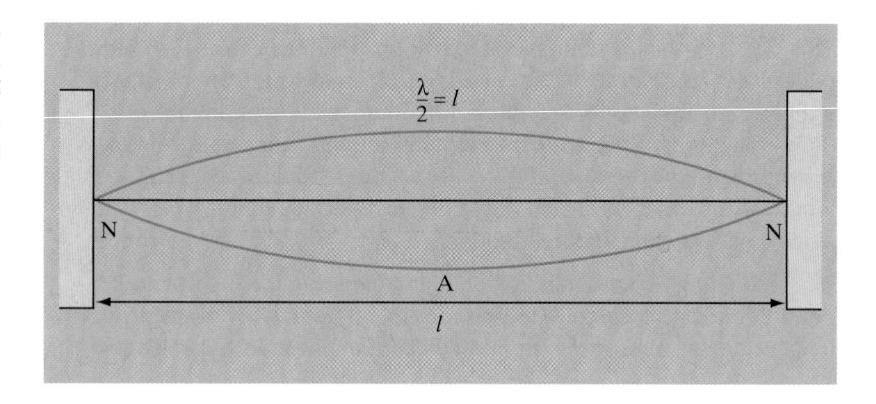

Fig 34
The fundamental mode of vibration for a string

N = node
A = antinode

The string can also support oscillations which have a node in the centre of the string. This is called the first overtone (or the second harmonic). There is now a whole wave on the string, so $\lambda = l$. The frequency of the first overtone is $f_1 = c/l$. This is twice the frequency of the fundamental.

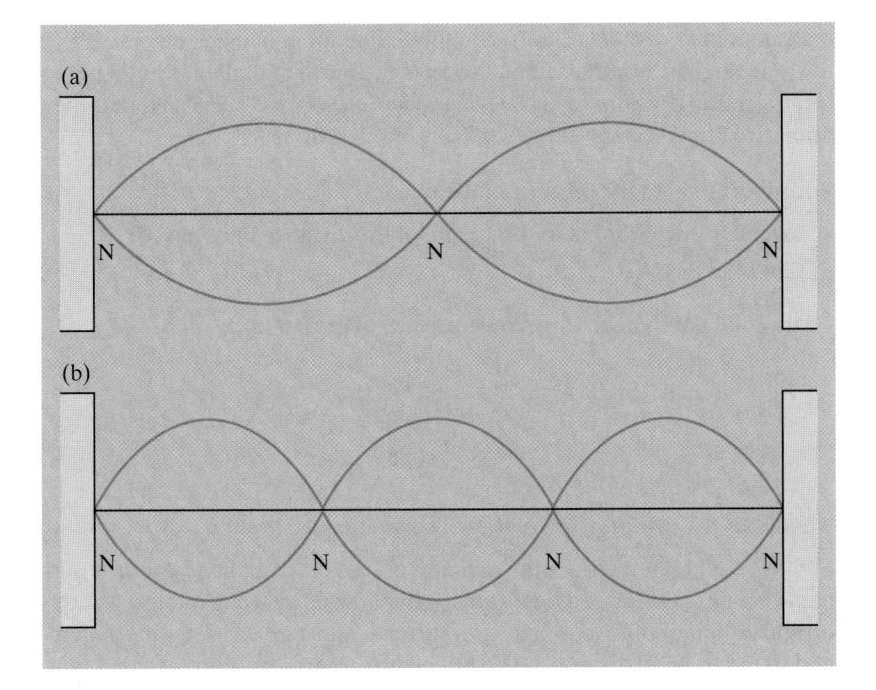

Fig 35
The overtones for a wave on a string

(a) first overtone (second harmonic)

(b) second overtone (third harmonic)

Example

(a) Standing waves of different frequencies can be produed on a cello string. Sketch the first and second harmonics.
(b) The strings on a cello are about 70 cm long. The lowest string vibrates at a frequency of 66 Hz. Assuming that this is the fundamental mode, calculate the speed of waves along the string.

Answers

(a)

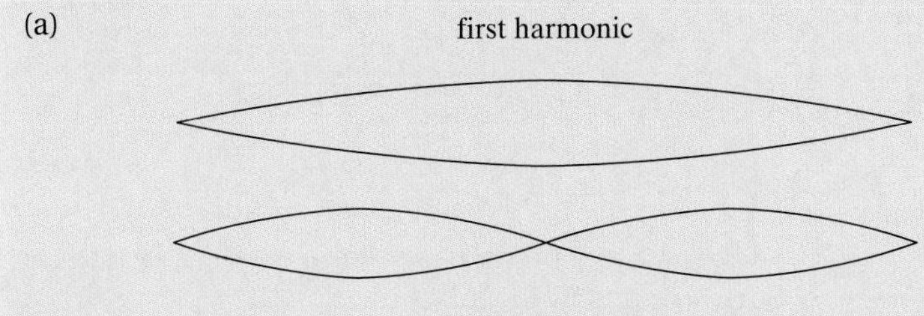

Fig 36

(b) The fundamental occurs when half a wavelength fits the string. $\lambda/2 = 0.7$ m, so $\lambda = 1.4$ m.

Speed of the wave $c = f\lambda = 66 \times 1.4 = 92.4\,\mathrm{m\,s^{-1}}$.

3.4.2 Gravitation

Newton's law

Gravity is the attractive force that acts between all masses. Isaac Newton realised that gravity was a universal force that not only made objects fall towards the Earth's surface, but also kept the Earth and planets in their orbits around the Sun. In his law of gravitation Newton suggested that the attractive force, F, between two masses, m_1 and m_2 is:

- proportional to the product of the masses, $F \propto m_1 m_2$;
- inversely proportional to the square of the distance between the masses, $F \propto 1/r^2$.

Examiners' Notes

This equation is given on the data sheet. Make sure you know what each symbol represents.

These rules combine to form **Newton's law of gravitation:**

$$F = \frac{G m_1 m_2}{r^2}$$

where G is the universal gravitational constant.

Newton could not measure the size of G and so was unable to confirm this equation directly. Instead he tested his hypothesis by using it to predict the orbital period of the Moon. The gravitational constant was later measured and found to be $6.67 \times 10^{-11}\,\mathrm{N\,m^2\,kg^{-2}}$. The reason why Newton could not measure G directly, and the reason why G is still known with less precision than many other constants, is that the gravitational force is quite weak. The force between two identical 1 kg masses placed 1 m apart is equal to:

$$F = \frac{G m_1 m_2}{r^2} = \frac{6.67 \times 10^{-11} \times 1 \times 1}{1^2} = 6.67 \times 10^{-11}\,\mathrm{N}$$

Fig 37
Definitions of m, r and F

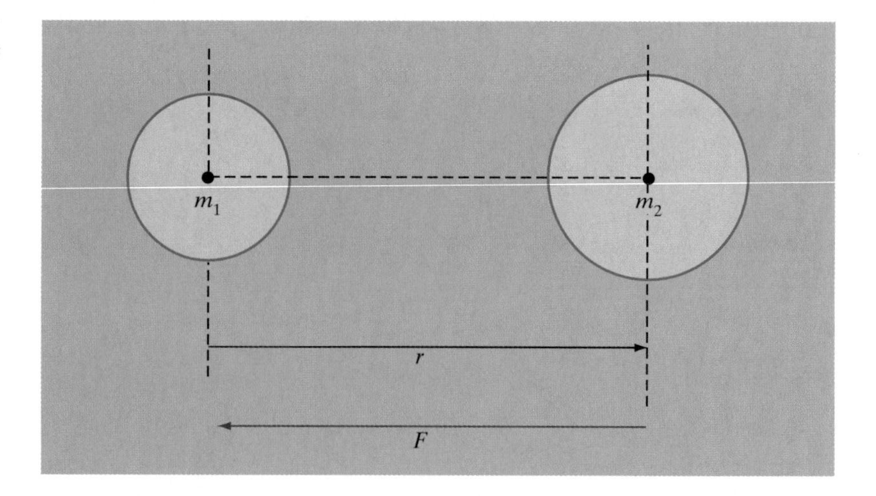

Example

Estimate the attractive gravitational force between two students sitting next to each other in a classroom.

Answer

Estimate the mass of the students as 60 kg each, and that there will be approximately 1 m between the centres of gravity of the students. If we apply Newton's law of gravitation:

$$F = \frac{Gm_1m_2}{r^2} = \frac{6.67 \times 10^{-11} \times 60 \times 60}{1^2} = 2.4 \times 10^{-7}\,\text{N}$$

This force acting on its own would cause one of the students to accelerate towards the other at around $4 \times 10^{-9}\,\text{m s}^{-2}$. No wonder you don't feel the attraction!

Gravitational field strength

Every mass causes a gravitational field around itself. A gravitational field is a region where any other mass will experience a force. The **gravitational field strength**, g, at a point is defined as the force that acts on a unit mass placed at that point. It is a vector quantity.

Definition

Gravitational field strength at a point is the force per unit mass, $g = F/m$.

The SI unit of gravitational field strength is N kg^{-1}.

A gravitational field can be represented by sketching field lines around a mass. The field lines show the direction of the force that acts on another mass. Near the Earth's surface the field lines are approximately parallel to each other and perpendicular to the surface of the Earth. The field here can be treated as uniform and the strength doesn't change significantly with height. On a larger scale the field lines around a spherical mass spread out radially.

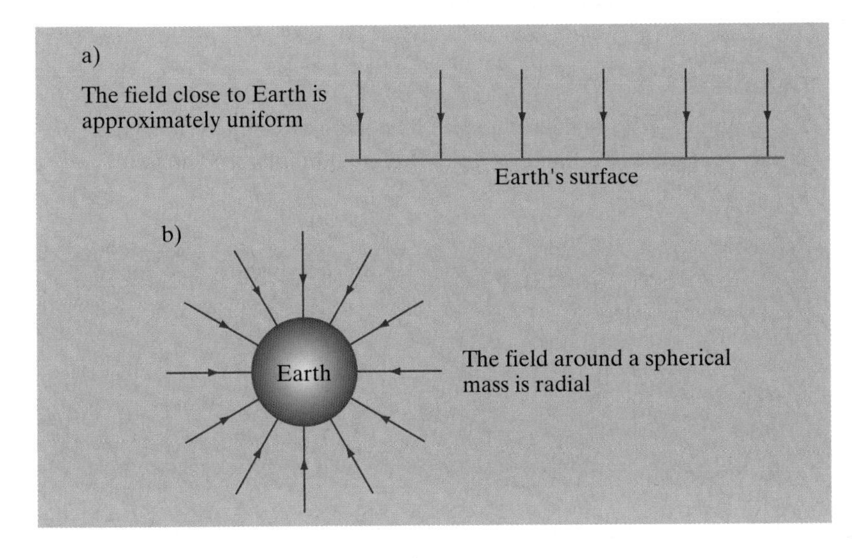

Fig 38
Field lines

Essential Notes

Gravitational field strength is a vector quantity. Its direction is that of the force acting on a mass in the field.

For a radial field, such as that due to a spherical mass M, the force, F, that acts on a small mass, m, placed at a distance, r, is given by Newton's law as $F = GMm/r^2$. The field strength is $g = F/m = GM/r^2$.

Definition

For a radial gravitational field, the magnitude of the field strength at a distance r from a mass M is $g = GM/r^2$.

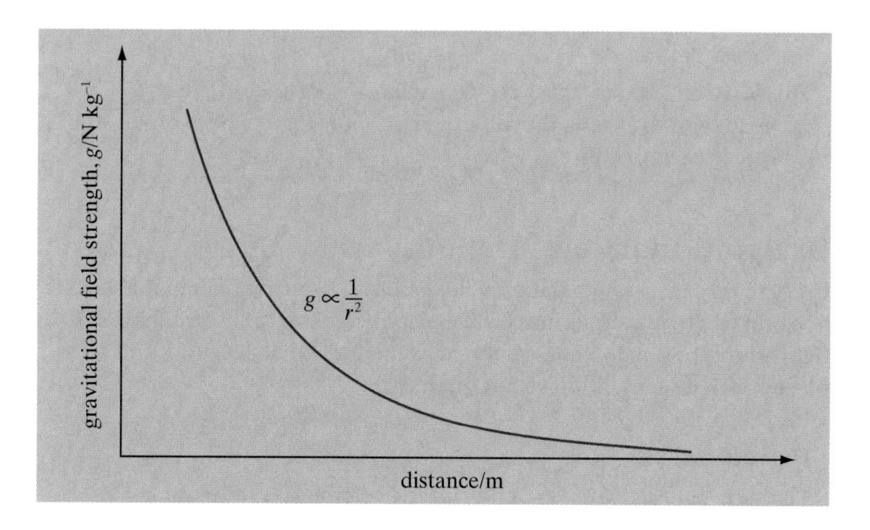

Fig 39
Gravitational field strength versus distance (radial field)

Gravitational potential

Consider a mass, m, moving in the Earth's gravitational field. As the mass moves closer to the Earth its potential energy changes. Usually we calculate the change in potential energy as $\Delta E_p = mg\Delta h$. This is acceptable for small values of Δh, but for larger values we need to take into account the fact that the gravitational field strength is not constant. In order to be able to calculate energy changes in a gravitational field we define a quantity called the **gravitational potential**.

Definition

The gravitational potential at a point in a gravitational field is defined as the work done in bringing a unit mass from infinity to that point.

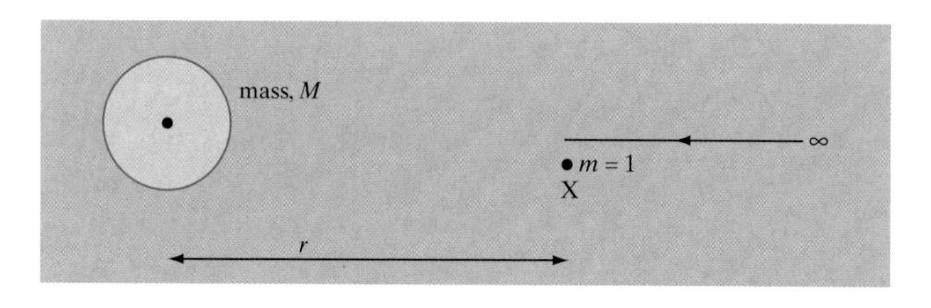

Fig 40
Defining gravitational potential

The gravitational potential at point X is equal to the work done in bringing a unit mass from infinity to the point

The work done, W, is defined as the force $\times$ the distance moved in the direction of the force.

The force on a unit mass at a distance r from a mass M is $F = GM/r^2$. So the work done in moving the mass through a small distance, Δr, is $-\Delta r \times GM/r^2$. The negative sign shows that the force, F, acts in the opposite direction to the displacement, Δr.

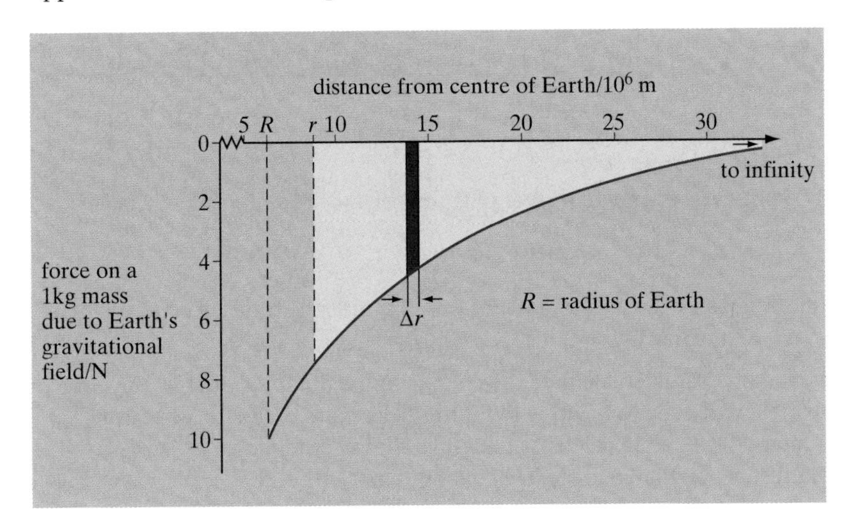

Fig 41
Work done in moving 1 kg mass from infinity to a distance r from Earth

The total work done in moving from infinity to a distance r is the sum of all strips that make up the shaded area. Each strip has an area of $F\Delta r = Gm\,\Delta r/r^2$. The sum of all such strips is equal to Gm/r.

Essential Notes

Gravitational potential is a scalar quantity; it has no associated direction.

Definition

The gravitational potential, V, at a distance r from a mass M is the work done in bringing a unit mass from infinity to the distance r. In a radial field V = −GM/r.

The negative sign indicates that this is a 'bound' system: energy would have be transferred to the unit mass in order to pull it away.

Examiners' Notes

Notice that the gravitational potential is proportional to $1/r$, whilst the gravitational field strength is proportional to $1/r^2$. If you are asked to sketch these graphs in an exam, try to show field strength falling away more steeply with distance than the potential does.

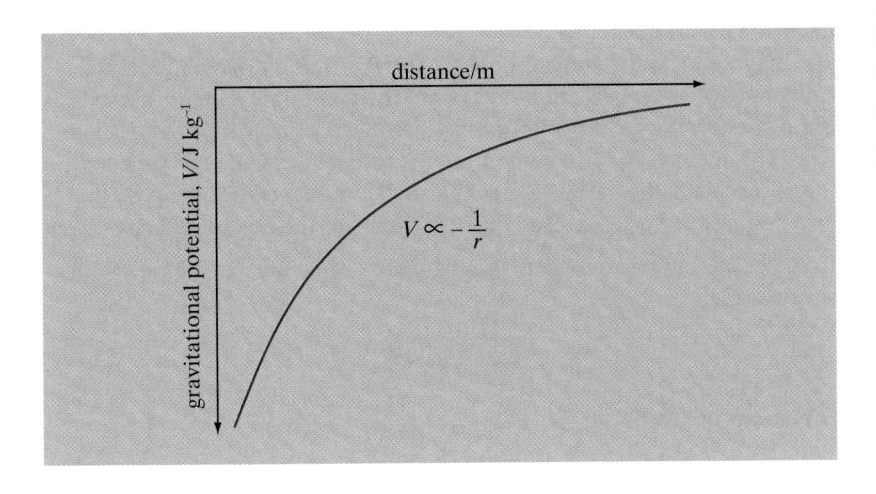

Fig 42
The variation of gravitational potential with distance (radial field)

Examiners' Notes

Don't confuse **potential** with **potential energy**. Potential, V, is a property of the field. It is used to describe the energy of a unit mass at a point in a field and is measured in joules per kilogram, J kg^{-1}. Potential energy, E_p, is calculated for a given mass, m, and has units of joules, J. The potential energy of a mass m which is placed in a field at a point with potential V is $E_p = mV$.

The work done in moving a mass in a gravitational field corresponds to the change in the gravitational potential energy of the mass. In other words, the change in gravitational potential can be thought of as the change in potential energy per unit mass. Gravitational potential has units of joules per kilogram, J kg^{-1}. To calculate the potential energy we need to multiply the potential by the mass that is in the field: $E_p = mV$.

The gravitational potential at the surface of the Earth is

$$V = -\frac{GM}{R} = -\frac{6.67 \times 10^{-11}\,\text{N m}^2\,\text{kg}^{-2} \times 5.98 \times 10^{24}\,\text{kg}}{6.37 \times 10^6\,\text{m}}$$

$$= -6.2616 \times 10^7\,\text{J kg}^{-1}$$

The potential energy of a 10 kg mass on the Earth's surface is

$$10 \times -6.2616 \times 10^7\,\text{J kg}^{-1} = -6.2616 \times 10^8\,\text{J}$$

In other words, an energy of 6.2616×10^8 J must be transferred to the mass to move it to infinity.

It is often useful to calculate the change in the potential energy. How much work would need to be done to lift the 10 kg mass to the top of Mount Everest? We need to calculate the gravitational potential at the top of Everest, a height above sea level of about 8800 m:

$$V = -\frac{GM}{r}$$

$$= -\frac{6.67 \times 10^{-11}\,\text{N m}^2\,\text{kg}^{-2} \times 5.98 \times 10^{24}\,\text{kg}}{(6.37 \times 10^6\,\text{m}) + 8800\,\text{m}}$$

$$= -6.2530 \times 10^7\,\text{J kg}^{-1}$$

Essential Notes

The gravitational potential has only changed a little, from sea level to the top of Everest. If we had assumed that the field was uniform, we could have used the expression $\Delta E_p = mg\Delta h$, which would have given the answer $10 \times 9.81 \times 8800 = 8.632 \times 10^5$ J, a difference of only 0.5%.

The difference in gravitational potential ΔV between sea level and the top of Everest is

$$(-6.2530 + 6.2616) \times 10^7 = 8.60 \times 10^4\,\text{J kg}^{-1}$$

So the energy required is $m\Delta V = 10 \times 8.60 \times 10^4 = 8.60 \times 10^5$ J.

Escape velocity

The magnitude of the gravitational potential at a point is the amount of energy needed to move a unit mass from there to infinity. In other words, it is the energy required to pull the mass completely out of the gravitational field. This energy could all come from the initial kinetic energy. We can therefore calculate how fast a mass has to be thrown vertically up, so that it never comes down again. This speed is known as the **escape velocity**.

If a mass has enough energy to escape then, neglecting work done against air resistance,

$$\Delta E_k = -\Delta E_p$$

For a mass on the Earth:

$$\tfrac{1}{2}mv^2 = \frac{GMm}{R}$$

So: $$\tfrac{1}{2}v^2 = \frac{GM}{R}$$

The escape velocity is therefore:

$$v = \sqrt{\frac{2GM}{R}}$$

For an object on the surface of the Earth where $g = 9.81\,\text{N kg}^{-1}$, $v = 11\,\text{km s}^{-1}$. If you could throw an object, of any mass, up at this speed, it would never come down.

Essential Notes

You don't have to reach escape velocity to leave the Earth. You could walk to the Moon if you had a long enough ladder. The escape velocity is the velocity required when all the energy has to come from the initial kinetic energy. Space rockets don't need to travel this fast, because they transfer energy in several bursts as they move further away from the Earth.

Gravitational field strength and gravitational potential

It is easy to get gravitational field strength and gravitational potential confused. Field strength is the *force* per unit mass, whereas potential is the *energy* per unit mass. The two quantities are connected.

In any gravitational field, the greater the field strength, the more the potential changes with distance.

Definition

The rate of change of potential with distance is equal to the field strength: $g = -\Delta V/\Delta x$.

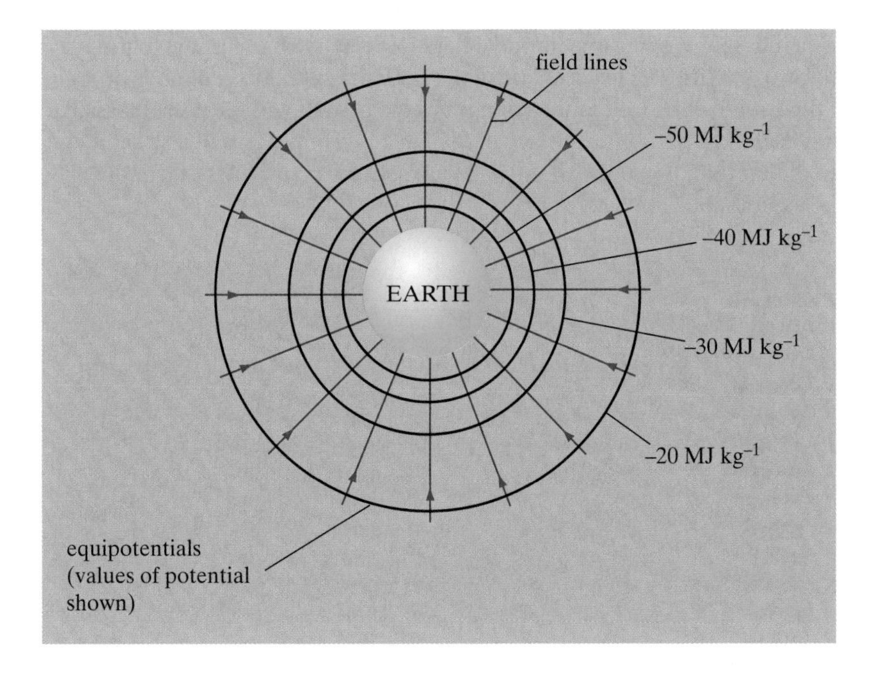

Fig 43
Field lines and equipotentials (surfaces of equal potential) around the Earth

Around a spherical mass the equipotentials are spherical. The field lines, showing the direction of the force, cross the equipotentials at right angles

Fig 44
Contour lines connect places of the same height, and therefore the same potential

Equipotentials are similar to contour lines on a map (Fig 44); they connect points where a mass would have the same potential energy.

Close to the Earth's surface, A in Fig 45, where the gravitational potential is changing quickly with distance, the gravitational field strength is strong. Further away, at B, $\Delta V/\Delta x$ is less, and so g is smaller.

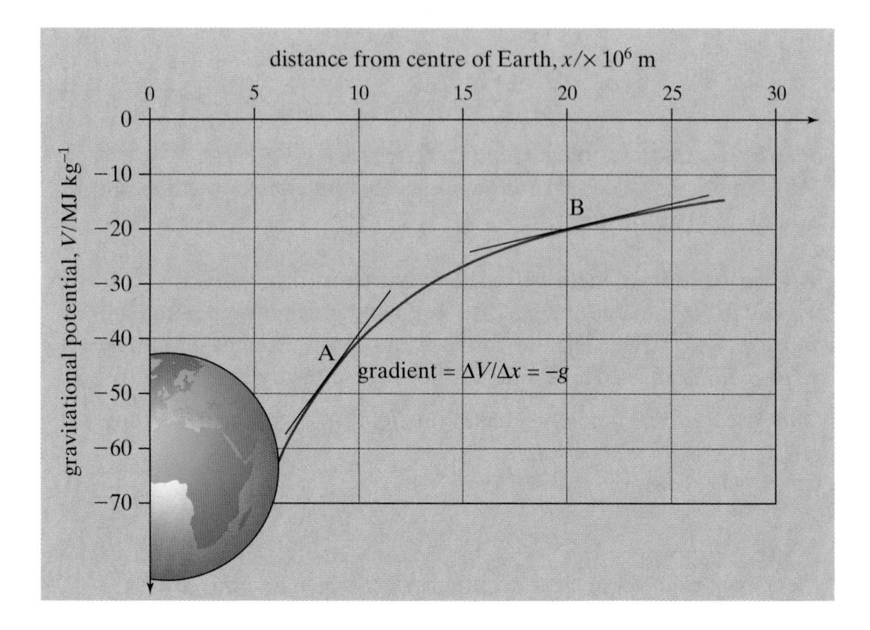

Fig 45
Potential of the Earth's gravitational field

Orbits of planets and satellites

As a mass moves in a gravitational field its velocity changes due to the force of gravity. The exact shape of the path that it follows depends on its initial velocity, as well as the strength of the field. If you release a mass, like a cricket ball, above the surface of the Earth, it simply falls in a straight line to the ground. If you throw the ball with some initial velocity, its path will be a parabola.

Fig 46
Motion under the influence of gravity

The motion of a mass in the Earth's gravitational field depends on its initial velocity

If we consider faster-moving objects, like a spacecraft approaching a planet, the general shape of the path is a hyperbola. At one particular speed the spacecraft can go into a circular orbit around the planet. This happens when the gravitational force equals the centripetal force necessary to make the spacecraft move in a circular path.

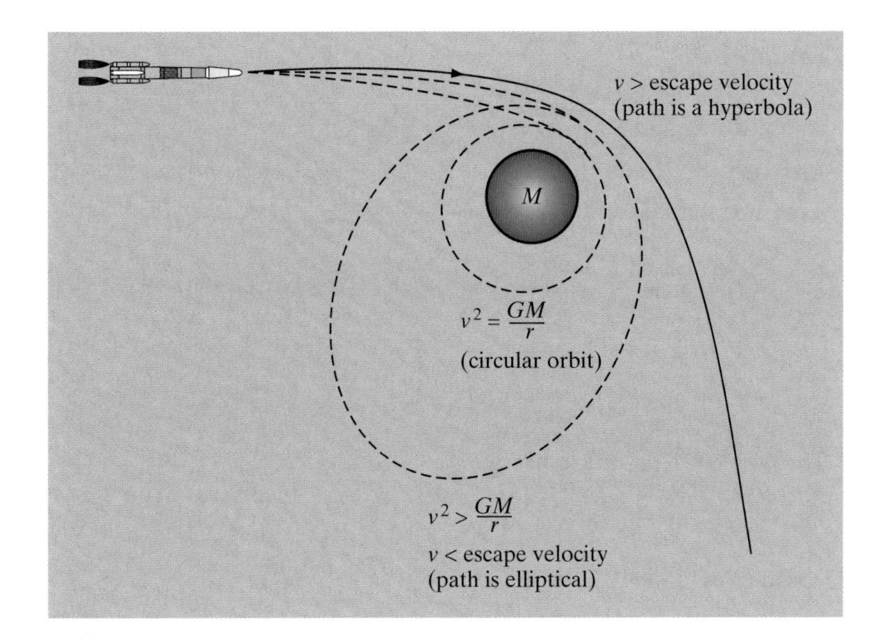

Fig 47
Motion in a gravitational field

Examiners' Notes

The general path of a satellite in an orbit, such as that taken by planets around the Sun, is elliptical. However, in this specification only circular orbits are studied.

When a satellite moves in a circular orbit, the centripetal force is provided by gravity. For a satellite of mass m moving in an orbit of radius r around a planet of mass M, this gives

$$F = \frac{GMm}{r^2} = \frac{mv^2}{r}$$

So

$$v^2 = \frac{GM}{r}$$

and

$$v = \sqrt{\frac{GM}{r}}$$

The orbital speed depends on the radius of the orbit and the mass of the planet, but not on the mass of the satellite.

For artificial satellites orbiting the Earth, those in a lower orbit have higher orbital speeds. Satellites used for Earth observations, such as weather satellites, typically have an orbital period of 90 minutes. Satellites used for relaying television pictures have to be in a fixed position relative to the Earth, so that domestic satellite dishes don't have to 'track' the satellite across the sky. These **geostationary** satellites have an orbital period of 24 hours and are placed in a very high equatorial orbit.

Example

Calculate the radius of a geostationary orbit. Take the Earth's mass as 5.98×10^{24} kg and G as 6.67×10^{-11} N m^2 kg^{-2}.

Answer

Assuming that the orbit is circular,

$$F = \frac{GMm}{r^2} = \frac{mv^2}{r}$$

which gives

$$\frac{GM}{v^2} = r \qquad (1)$$

The time taken for one orbit is

$$T = \frac{\text{distance}}{\text{speed}} = \frac{2 \times \pi \times r}{v}$$

which gives

$$v = \frac{2\pi r}{T} \qquad (2)$$

Combining equations (1) and (2) to eliminate v,

$$\frac{GM}{(2\pi r/T)^2} = r \quad \text{and so} \quad \frac{GMT^2}{4\pi^2} = r^3$$

The orbital period has to be 24 hours: $T = 24 \times 60 \times 60 = 86\,400$ s

$$\text{So } r^3 = \frac{6.67 \times 10^{-11} \times 5.98 \times 10^{24} \times 7.465 \times 10^{9}}{39.48} = 7.54 \times 10^{22}\,\text{m}^3$$

This gives a value for r of 42.2×10^6 m.

Since the Earth's radius is 6.38×10^6 m, a geostationary satellite has to be placed in an orbit that is approximately 6 Earth radii above the Earth's surface.

The energy of an orbiting satellite

The total energy of a satellite is the sum of its potential energy and its kinetic energy. If the satellite is moving at a velocity v in a circular orbit of radius r around the Earth then:

$$E_{\text{total}} = E_k + E_p = \tfrac{1}{2}mv^2 - \frac{GMm}{r}$$

For a circular orbit, equating the gravitational force to the centripetal force,

$$\frac{mv^2}{r} = \frac{GMm}{r^2}$$

So, substituting,

$$E_{\text{total}} = \frac{GMm}{2r} - \frac{GMm}{r} = -\frac{GMm}{2r}$$

We can apply this to a satellite that is spiralling towards Earth, losing energy as it enters the outer part of the Earth's atmosphere. As it gets closer to Earth, r decreases and E_{total} becomes more negative (the total energy decreases). However, the decrease in total energy is only half the decrease in potential energy. The difference is due to the increase in kinetic energy. Rather strangely, the satellite speeds up as it loses energy, eventually burning up or crashing onto the Earth's surface at high speed.

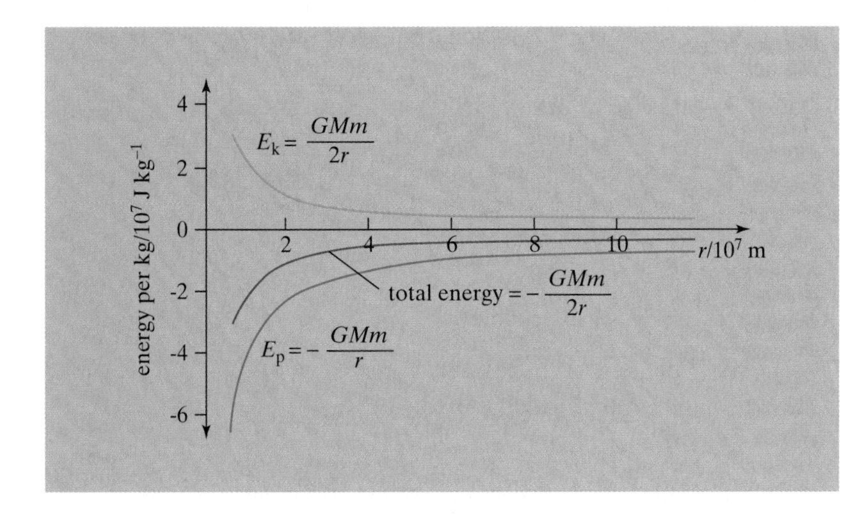

Fig 48
Orbital energy

3.4.3 Electric fields

Coulomb's law

There is an electrostatic force that acts between all electric charges. This force is described by **Coulomb's law**. Coulomb's law states that the force, F, between two charges, Q_1 and Q_2, which are separated by a distance r is:

- proportional to the product of the two charges, $F \propto Q_1Q_2$;
- inversely proportional to the square of the distance between the charges, $F \propto 1/r^2$.

This force law is similar to Newton's law of gravitation (page 38). Both laws follow the inverse-square relationship. One difference is that the force between two charges depends on the medium between the charges, whereas the gravitational force is independent of the medium. For two charges in a vacuum the force between them is

$$F = \frac{Q_1Q_2}{4\pi\varepsilon_0 r^2}$$

The constant ε_0 is known as the **permittivity of free space**. The value of ε_0 is $8.85 \times 10^{-12}\,\mathrm{F\,m^{-1}}$. The force between two 1 coulomb charges placed 1 m apart in a vacuum is therefore

$$F = \frac{1}{4\pi\varepsilon_0} = 9 \times 10^9\,\mathrm{N}$$

Essential Notes

If the two charges are separated by another medium, such as air, the force between them is reduced and the equation becomes

$$F = \frac{Q_1Q_2}{4\pi\varepsilon r^2}$$

ε is the permittivity of the material, which is usually given in terms of how much bigger it is than the permittivity of free space, $\varepsilon = \varepsilon_r\varepsilon_0$, where ε_r is the relative permittivity.
ε_r for air is 1.005, so we can usually neglect the effect of air.

The electrostatic force can be attractive or repulsive. Two charges of similar sign, two positives or two negatives, will repel each other. The force between two opposite charges, positive and negative, will be attractive. An attractive force will have a negative sign.

Example

A hydrogen atom consists of a single electron orbiting a single proton. Compare the size of the gravitational and electrostatic forces between the particles.

Data:		
	Mass of proton	1.67×10^{-27} kg
	Mass of electron	9.11×10^{-31} kg
	Charge on an electron	-1.60×10^{-19} C
	Charge on a proton	1.60×10^{-19} C
	Radius of a hydrogen atom	5.29×10^{-11} m

Answer

The size of the electrostatic force is

$F = Q_1Q_2/4\pi\varepsilon_0 r^2 = 8.23 \times 10^{-8}$ N

The size of the gravitational force is

$F = G\,m_1m_2/r^2 = 3.63 \times 10^{-47}$ N

The electrostatic force between the proton and the electron is about 2×10^{39} times larger than the gravitational force.

Electric field strength

Every charged particle creates an electric field around itself. An electric field is a region of space where a charged particle experiences a force. We can visualise electric fields by sketching the field lines around a charge. The field lines show the direction of the force that would be exerted on a positive charge placed in the field.

Around a point charge, or a spherical charge distribution, the field lines are radial. See Fig 49, opposite.

The strength of the electric field, E, is defined as the force, F, that would be exerted on a unit charge. Field strength is measured in newtons per coulomb, N C^{-1}.

Definition

Electric field strength, E, *is the force per unit charge:* E = F/Q.

Essential Notes

Electric field strength is a vector quantity. Its direction is that of the force acting on a positive charge.

Since the force between two point charges, Q and q, is $F = Q\,q/4\pi\varepsilon_0 r^2$, the electric field strength in a radial field is

$$E = \frac{F}{q} - \frac{Q}{4\pi\varepsilon_0 r^2}$$

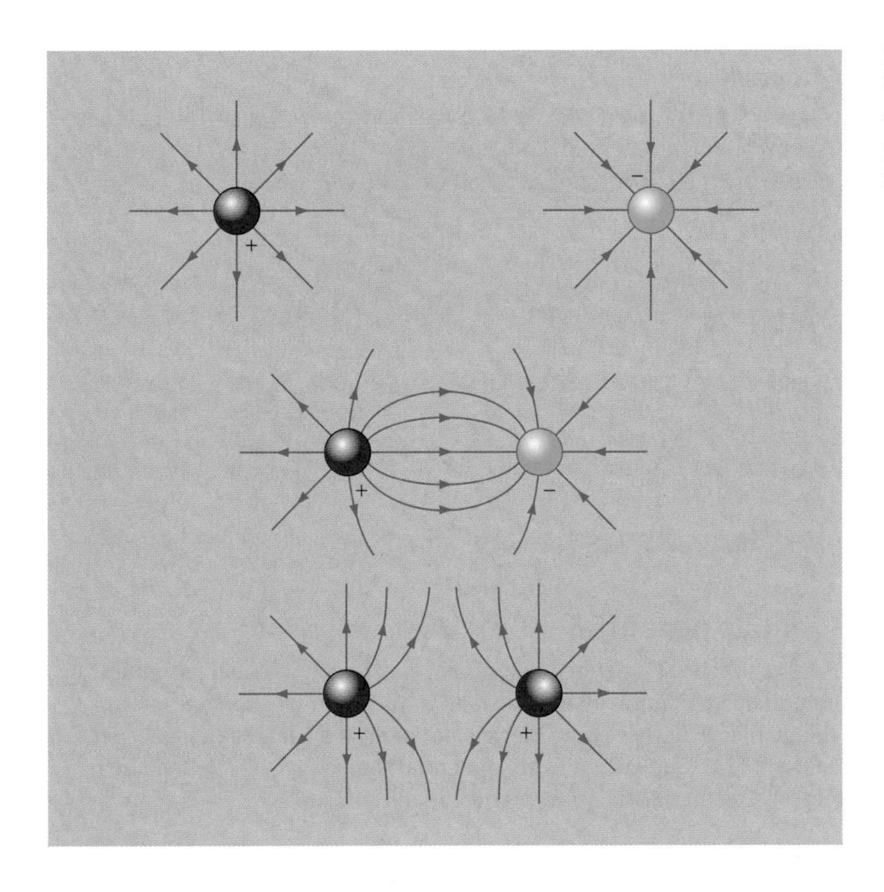

Fig 49
Field lines around charges

The field lines point away from a positive charge, but go in towards a negative charge

Uniform field

It is possible to create a uniform electric field between two oppositely charged parallel conductors, such as the plates of a capacitor. In between the plates the field strength is independent of position and just depends on the separation of the plates, d, and the potential difference, V, between them.

The electric field strength in a uniform field is

$$E = \frac{V}{d}$$

Examiners' Notes

Electric field strength can be measured in units of $\mathrm{N\,C^{-1}}$ or $\mathrm{V\,m^{-1}}$.

Essential Notes

There is no electric field inside a closed conductor. A hollow metal container, known as a Faraday cage, may have charges on the outside but will not have any charges on its inside surface. Faraday cages are used to shield people or sensitive equipment from intense electric fields.

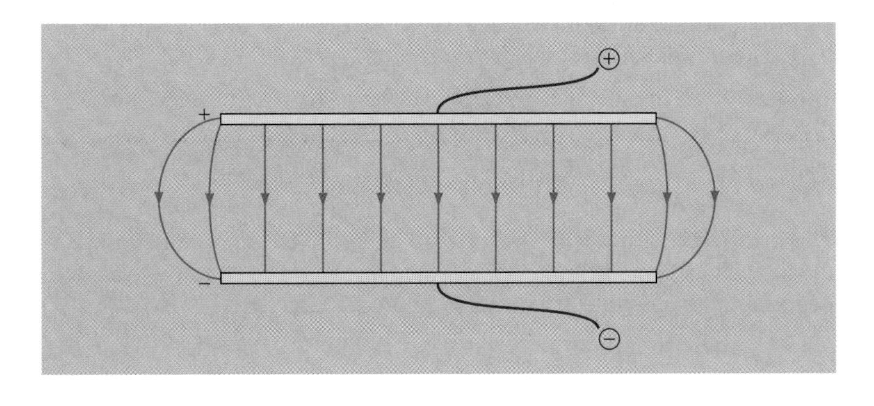

Fig 50
A uniform electric field

Example

An electron is placed between two metal plates, which are 10 cm apart in a vacuum. A potential difference of 100 V is applied across the plates. Calculate the magnitude of the force on the electron.

Answer

The electric field strength in this uniform field is

$$E = \frac{V}{d} = \frac{100}{0.1} = 1000\,\mathrm{V\,m^{-1}}$$

Field strength is the force on a unit charge, so the force on a charge q is $F = qE$.

The charge on an electron is -1.6×10^{-19} C. So the magnitude of the force on the electron is

$F = 1.6 \times 10^{-19} \times 1000 = 1.6 \times 10^{-16}\,\mathrm{N}$

Electric potential

Electric potential is defined in a similar way to gravitational potential. The potential at a point in an electric field is the potential energy of a unit charge placed at that point. Electric potential is measured in joules per coulomb, $\mathrm{J\,C^{-1}}$, or volts. Electric potential can be formally defined in a similar way to gravitational potential (page 38).

Definition

*The **electric potential,** V, at a point in an electric field is defined as the work done in bringing a unit positive charge from infinity to that point.*

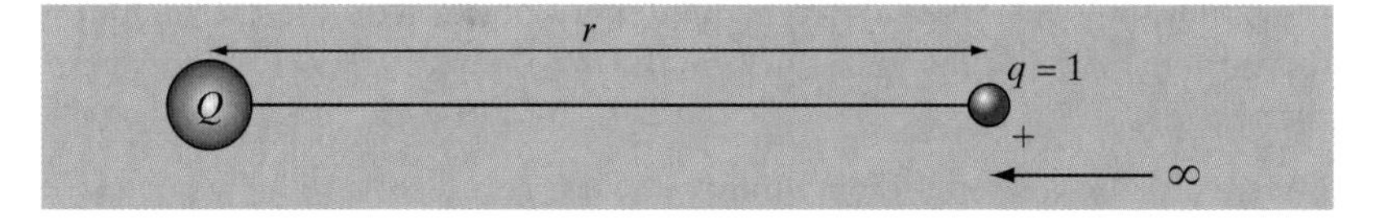

Fig 51
The definition of potential

Essential Notes

Electric potential is a scalar quantity; it has no associated direction.

The work done in moving against a force is given by force × distance. In a radial electric field the force on a charge, q, changes with distance, r, so we cannot use this formula directly. Instead we need to consider the work done in moving a small distance, Δr, because the force can be assumed to be constant over such a small distance.

The work done in moving a unit positive charge ($q = 1$) through a small distance Δr is

$$\Delta W = F\Delta r = Q\Delta r/4\pi\varepsilon_0 r^2$$

This is the area of the small shaded strip in Fig 52. The total work done is the total area below the force curve. This gives $W = Q/4\pi\varepsilon_0 r$, which is, by definition, the potential at a distance r from a charge Q.

The electric potential at a distance r from a charge Q is

$$V = \frac{Q}{4\pi\varepsilon_0 r}$$

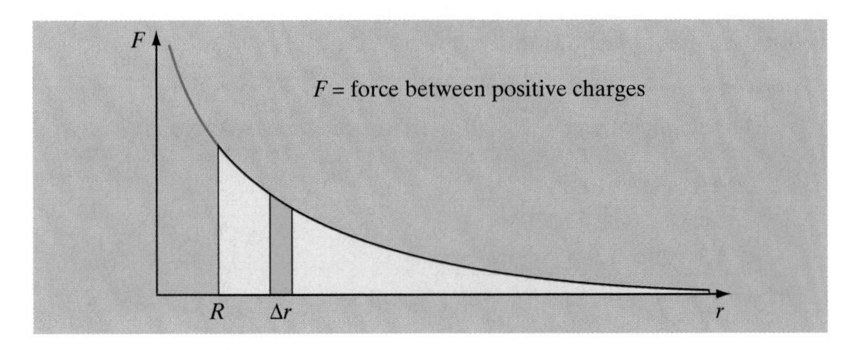

Fig 52
Work done in an electric field

The total work done in bringing a charge from infinity to R is the area under the curve

The **potential difference** between two points, A and B, is the work done in moving a unit positive charge from B to A. In SI units charge is measured in coulombs, work done is measured in joules, and potential difference is measured in volts. There is a potential difference of 1 volt between A and B when one joule of work is done in moving one coulomb of charge from B to A.

In general the work done, W, in moving a charge, q, through a potential difference V is $W = qV$. (You have met this relationship before, in Unit 1.)

The potential at different points of a conductor is always the same. If there were any potential differences, charge would flow until the differences were evened out.

There is a positive potential in the region of a positive charge (Fig 53). Work needs to be done to bring a positive charge from infinity to the region.

There is a negative potential in the region of a negative charge (Fig 53), since the value of Q in the equation $V = Q/4\pi\varepsilon_0 r$ is negative. A negative potential means that work would have to be done on a positive charge to pull it back to infinity. The situation is similar to the potential in a gravitational field, which is always negative ($V = -GM/r$).

Essential Notes

Suppose we move a charge of just one electron, rather than a whole coulomb of charge, through a potential difference of 1 volt. The work done then is

$W = qV = 1.6 \times 10^{-19}$ J

This small amount of energy is known as an **electron volt, eV**.

Examiners' Notes

Potential is inversely proportional to r, whereas field strength is inversely proportional to the *square* of r.

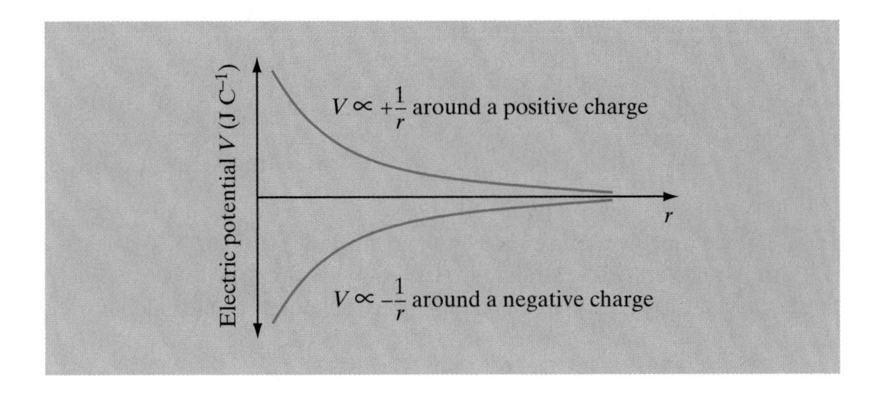

Fig 53
Electric potential in the field around a positive and negative charge

Essential Notes

Potential drops to zero at infinity, but infinity is not a very practical reference point. We often use the Earth's surface as zero potential. The Earth is a reasonably good conductor and is therefore all at the same potential.

Example

In Rutherford scattering an alpha particle approaches a gold nucleus. Calculate the work done by the alpha particle as it moves from a large distance away to within 0.5×10^{-12} m from the gold nucleus. (The proton number of gold is 79.)

Answer

The electric potential at 0.5×10^{-12} m from the gold nucleus is

$$V = \frac{Q}{4\pi\varepsilon_0 r}$$

$$= \frac{79 \times 1.6 \times 10^{-19}}{4\pi \times 8.85 \times 10^{-12} \times 0.5 \times 10^{-12}}$$

$$= 2.27 \times 10^5\,\mathrm{J\,C^{-1}}$$

Since the charge on the alpha particle is $2 \times 1.6 \times 10^{-19}$ C, the work done is

$$W = qV = 2 \times 1.6 \times 10^{-19} \times 2.27 \times 10^5 = 7.27 \times 10^{-14}\,\mathrm{J}$$

(Supposing this energy comes from the kinetic energy of the alpha particle, the particle would have to be emitted with an initial energy of $7.27 \times 10^{-14}\,\mathrm{J} = 0.45\,\mathrm{MeV}$.)

Essential Notes

Equipotentials in electric fields are similar to those in gravitational fields (page 41). They connect points where a charge has the same potential energy.

Motion of charged particles in an electric field

In a uniform electric field, charged particles accelerate in a similar way to masses moving in a gravitational field. In a uniform field the force, F, on a charge, q, is constant: $F = qE$. The direction of the force on a charge is also constant, with the result that the particle will move in a parabolic path.

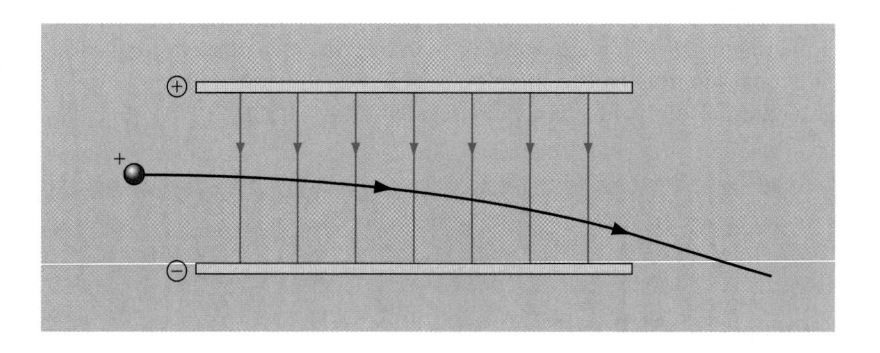

Fig 54
Charged particle moving in a field

A positive charge is attracted towards the negative electrode

For a particle moving along a field line, the motion is simpler. The particle will simply accelerate in a straight line, like a mass dropped towards the surface of the Earth. We can use the potential difference to calculate the velocity change. Imagine an electron accelerating through a potential difference of 250 V. The electron's potential energy will be transferred as kinetic energy:

$$qV = \frac{1}{2}mv^2$$

so

$$v = \sqrt{\frac{2qV}{m}}$$

For a potential difference $V = 250\,\text{V}$, this works out to be $9.38 \times 10^6\,\text{m s}^{-1}$.

Examiners' Notes

Treat these calculations with caution. As the potential gets higher, this equation would lead to electron speeds that are faster than the speed of light! A relativistic method for calculating speed is given in the optional Unit 5D.

Comparison of electric and gravitational fields

Gravitational and electric fields both have force laws which obey the inverse-square law. The major differences arise because there are positive and negative charges, but only positive masses. The gravitational force is always attractive, whereas the electric force may be attractive or repulsive. The gravitational force is independent of the medium between the masses, whereas the force between two charges is affected by the material between them.

Table 2
Gravitational force and electrostatic force compared

	Gravitational force	Electric force
Force law	Newton's law of gravitation $F = G\,m_1m_2/r^2$	Coulomb's law $F = Q_1Q_2/4\pi\varepsilon r^2$
Force proportional to	Product of masses $1/r^2$	Product of charges $1/r^2$
Constant of proportionality	G This is a universal constant. The force is independent of the medium	$1/4\pi\varepsilon$ The force depends on the permittivity of the intervening medium
Type of force	Always attractive	Can be attractive or repulsive
Field strength (radial)	$g = GM/r^2$ Force on unit mass, N kg^{-1}	$E = Q/4\pi\varepsilon r^2$ Force on unit charge, N C^{-1}
Potential (radial field)	$V = -GM/r$ Potential energy per unit mass, J kg^{-1}	$V = Q/4\pi\varepsilon_0 r$ Potential energy per unit charge, J C^{-1}

3.4.4 Capacitance

Capacitance

Capacitors are devices that store electric charge. At their simplest they consist of two metal plates separated by an air gap. When a capacitor is connected to a battery, negative charges (electrons) flow on to one plate and off the other plate. As the plates become charged, they oppose further charges flowing on to the plates and the current in the circuit drops. Eventually the potential difference across the capacitor is equal to the potential difference of the battery and no more current flows. The capacitor is now fully charged, with a charge of $+Q$ on one plate and a charge of $-Q$ on the other plate. It would be possible to store more charge on the capacitor by increasing the potential difference across the capacitor.

Fig 55
Capacitor in a circuit

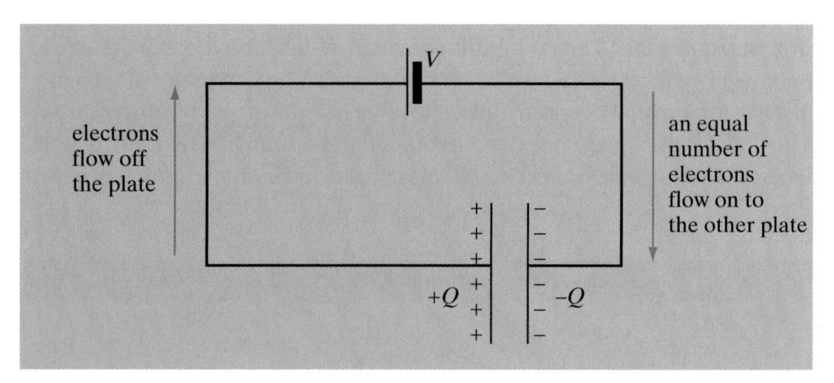

The charge stored, Q, is proportional to the potential difference, V, across the capacitor:

$$Q \propto V$$

We can write this as

$$Q = CV$$

where C is a constant known as the **capacitance**.

Fig 56
Charge stored on a capacitor versus potential difference applied

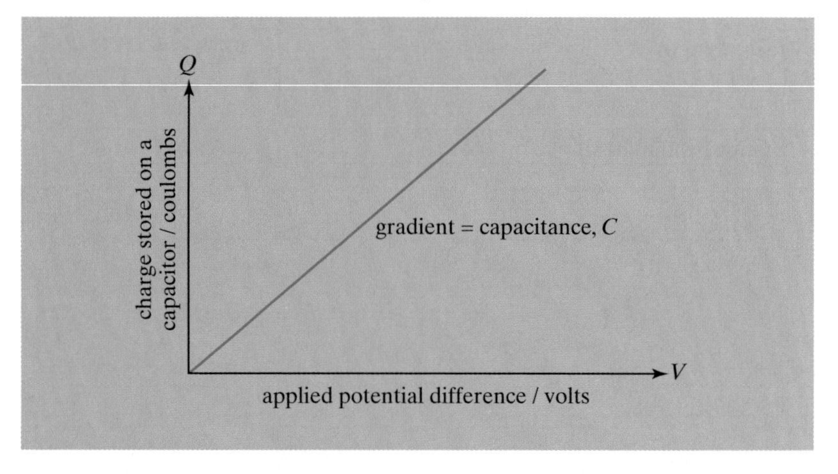

The capacitance is a constant for a given capacitor and its value depends on the construction of the capacitor, in particular the area and separation of the plates, and the material used between the plates.

If we rearrange the equation $Q = CV$, we can define capacitance: $C = Q/V$, as the charged stored per unit potential difference. The unit of capacitance is known as the **farad**.

Examiners' Notes

The unit of capacitance, the farad, is very large. Real capacitors tend to have values measured in millifarads, mF, microfarads, μF, nanofarads, nF, or picofarads, pF.

Definition

Capacitance is the charge stored by a capacitor, per volt of potential difference applied across it. A capacitance of one farad will store one coulomb of charge for every volt applied across the capacitor.

It is not possible to store more and more charge by applying an ever-increasing potential difference across the capacitor. Eventually the potential difference becomes too large for the insulating material between the plates. Electrical breakdown then occurs and conduction takes place across the gap. A capacitor is often labelled with its capacitance, in farads, and the maximum potential difference, in volts.

Essential Notes

The insulating material between the plates is referred to as the **dielectric**. Various materials are used to increase the capacitance. The type of material used, and its thickness, determine the maximum working potential for the capacitor.

Example

A capacitor is labelled 100 nF 10 V. Calculate the maximum charge that can be stored on the capacitor.

Answer

The maximum potential difference that can be applied across the capacitor is 10 V. The maximum charge stored will then be

$$Q = CV = 100 \times 10^{-9}\,\text{F} \times 10\,\text{V} = 1 \times 10^{-6}\,\text{C}$$

Energy stored by a capacitor

A charged capacitor is a store of electrical energy. The amount of energy stored depends on the charge stored by the capacitor and on the potential difference across it. The amount of energy stored can be found by considering the charging process.

When a capacitor is connected to a battery (as in Fig 55), charge flows on to the capacitor. The battery has to transfer energy to do work against the repulsive force from the charge that is already on the capacitor. The work done, W, in moving a charge Q through a potential difference V is $W = QV$. However, the potential difference across a capacitor during charging is not constant. The potential difference increases from zero, when the capacitor is uncharged, to V when it is fully charged. The average potential difference across the capacitor is therefore

$$\frac{V + 0}{2} = \tfrac{1}{2}V$$

Examiners' Notes

Remember the definition of potential difference from Unit 1: the potential difference is the energy transferred per unit charge. Since $V = E/Q$, energy (or work done) is given by $E = QV$.

The work done in charging a capacitor is then:

$$W = Q \times \tfrac{1}{2}V = \tfrac{1}{2}QV$$

and the energy stored by the capacitor is therefore:

$$E = \tfrac{1}{2}QV$$

Since $Q = CV$, the energy equation can be written in other forms:

$$E = \tfrac{1}{2}QV = \tfrac{1}{2}CV^2 = \tfrac{1}{2}\frac{Q^2}{C}$$

The energy stored can also be derived from a graph of potential difference against charge for the capacitor while charging. See Fig 57.

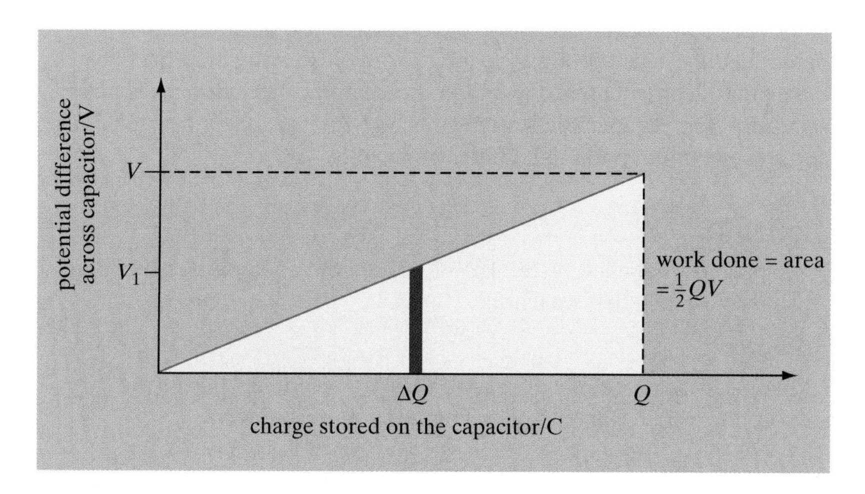

Fig 57
Energy stored by a capacitor

The work done in adding a small amount of charge ΔQ to the capacitor at a potential of V_1 is $\Delta W = \Delta Q \times V_1$. This is the area of the strip marked in the diagram. Adding together all the strips gives the work done in charging the capacitor fully. This is the area under the line, which gives $W = \tfrac{1}{2}QV$

Example

The capacitor used to store charge in a camera flash unit has a capacitance of 470 mF and can be charged to a potential of 30 V.

(a) How much energy is stored by the capacitor when it is fully charged?

(b) If the capacitor discharges through the flash in a time of 0.20 ms, calculate the average power.

Answer

(a) The energy stored is $E = \tfrac{1}{2}QV = \tfrac{1}{2}CV^2 = 0.5 \times 470 \times 10^{-3} \times 30^2$
$= 211.5\,\text{J}$.

(b) Power is rate of energy transfer $= 211.5/0.20 \times 10^{-3} = 1.06\,\text{MW}$.

Capacitor discharge

Charging a capacitor through a resistor

When an uncharged capacitor is placed in a circuit (see Fig 58) to be charged, the initial charging current, I_0, depends on the e.m.f. of the battery, E, and the resistance of the circuit, R. Since there is no charge on the capacitor to start with, the initial charging current I_0 is given by $I_0 = E/R$.

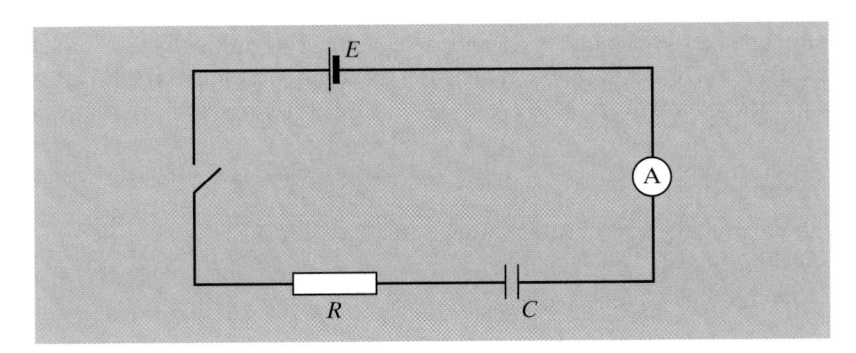

Fig 58
A charging circuit for a capacitor

The resistance R represents the total resistance of the circuit, including the internal resistance of the battery and the resistance of any connections

As charge accumulates on the capacitor, the potential difference across it increases. This potential difference opposes that of the battery and it becomes increasingly difficult to push charge on to the capacitor and the charging current drops. Eventually the potential difference across the capacitor is equal, but opposite, to that of the battery and the charging current drops to zero.

If we take readings of the charging current against time, we find that the current drops in an **exponential decay** curve (Fig 59).

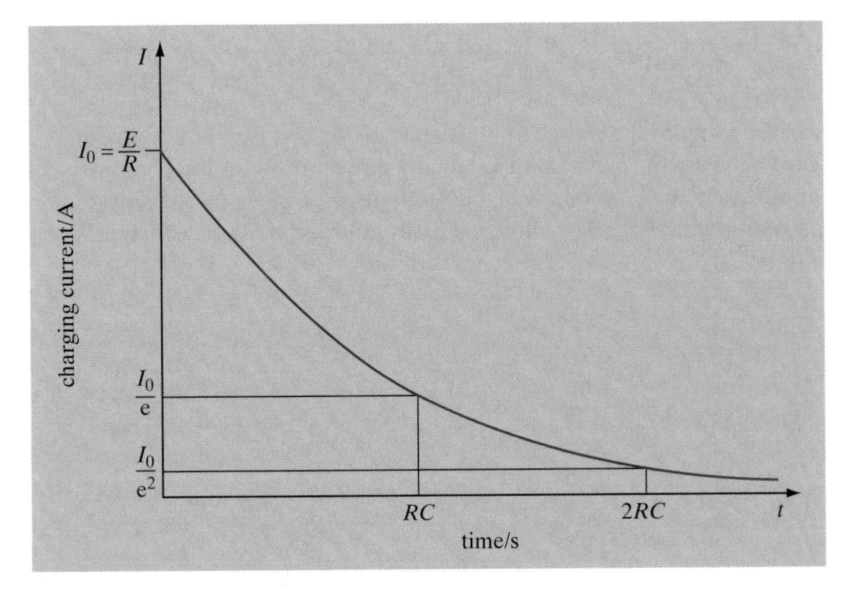

Fig 59
Charging current against time

Essential Notes

The equation for an exponential decay curve has the form $y = Ae^{-bx}$, where A and b are constants. You can find the value of the number e by entering e^1 on your calculator. You should get a value of 2.718281828 (to 10 s.f.).

The rate at which the current drops depends on the resistance, R, and on the capacitance, C, of the circuit. The larger the resistance, the lower the current. The capacitor therefore takes longer to charge. A large value of capacitance also increases the charging time, since it takes more charge to fill the capacitor.

The product RC is known as the **time constant** of the circuit.

Definition

The time constant is the time that it takes for the current to drop to 1/e (≈ 0.37) of its original value.

Examiners' Notes

Exponential decay curves have the property that they always take the same time to decrease by a given fraction.

After one time constant, $t = RC$ and the charging current will have dropped to $1/e$ of its original value. After two time constants, $t = 2RC$, the current will have dropped to $1/e^2$ (≈ 0.135 of the original value).

Example

A 100 μF capacitor is charged from a 10 V supply in a circuit of resistance 1 kΩ.

(a) What will the initial current be?

(b) What is the time constant of the circuit?

(c) What would the current be after a time equal to four time constants has passed?

Answer

(a) $I_0 = E/R = 10/1000 = 0.01\,A = 10\,mA$.

(b) Time constant $= RC = 1000 \times 100 \times 10^{-6} = 0.1\,s$.

(c) After four time constants, $t = 0.4\,s$, the current will have dropped to $1/e^4$ of its original value:

$1/e^4 = 0.0183$

So the current will be $0.0183 \times 10\,mA = 0.183\,mA$.

We can also plot the charge on the capacitor against time as it is charged. The charge on the capacitor is originally zero and rises quickly at first, since the charging current is highest at time $t = 0$. Since the current decreases with time, the charge gradually approaches its steady state value of $Q = CV$.

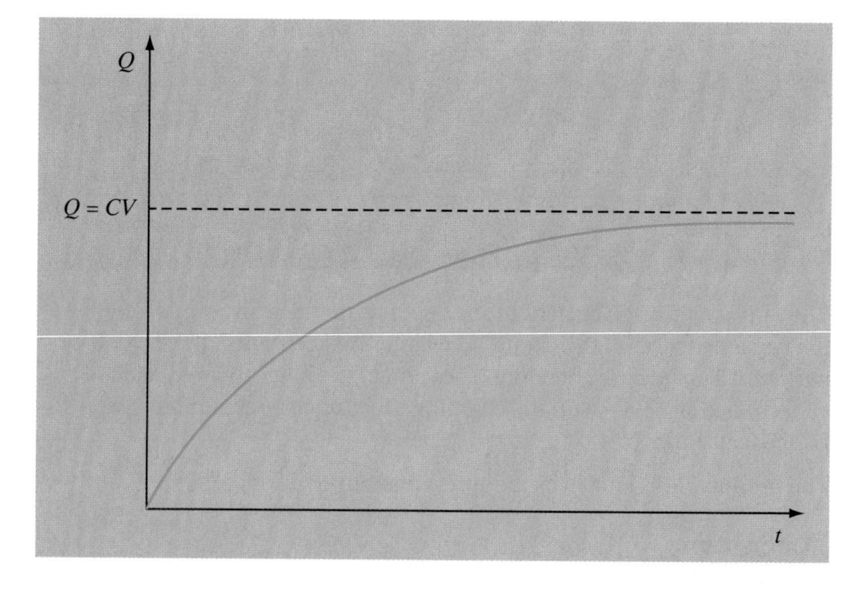

Fig 60
Charge versus time for a capacitor being charged

Essential Notes

Current is rate of change of charge, $\Delta Q/\Delta t$, so at any time the current is the gradient of this curve.

Since the potential difference across the capacitor is proportional to the charge, a voltage versus time graph has the same shape as Fig 60.

For example, in the circuit of Fig 61, when the switch is closed the capacitor charges up and so the potential difference measured by the voltmeter goes up in proportion.

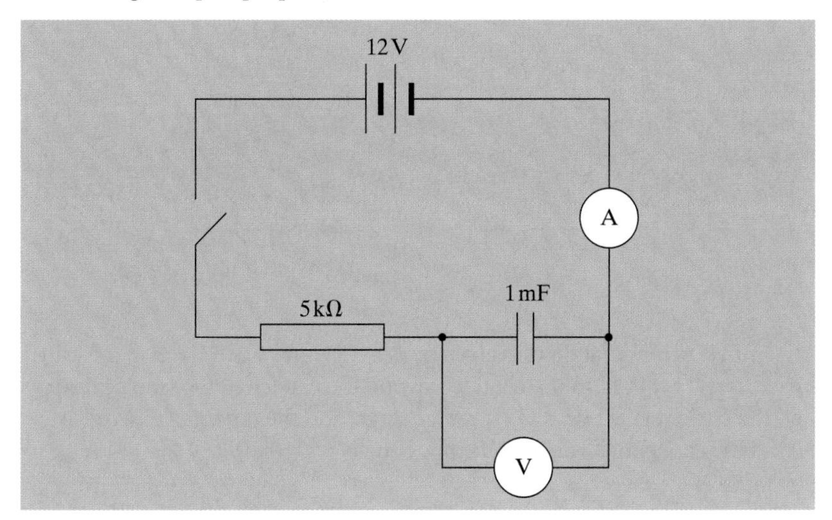

Fig 61

If the voltmeter was connected across the resistor instead of across the capacitor, it would initially show a maximum value, since $V = IR$ and the current will be at a maximum to start with. Eventually, the potential difference across the resistor will drop to zero (Fig 62). At any time the sum of the potential differences across the capacitor and the resistor will be equal to the e.m.f. supplied by the battery, in this case 12 V.

$$V_C + V_R = E$$

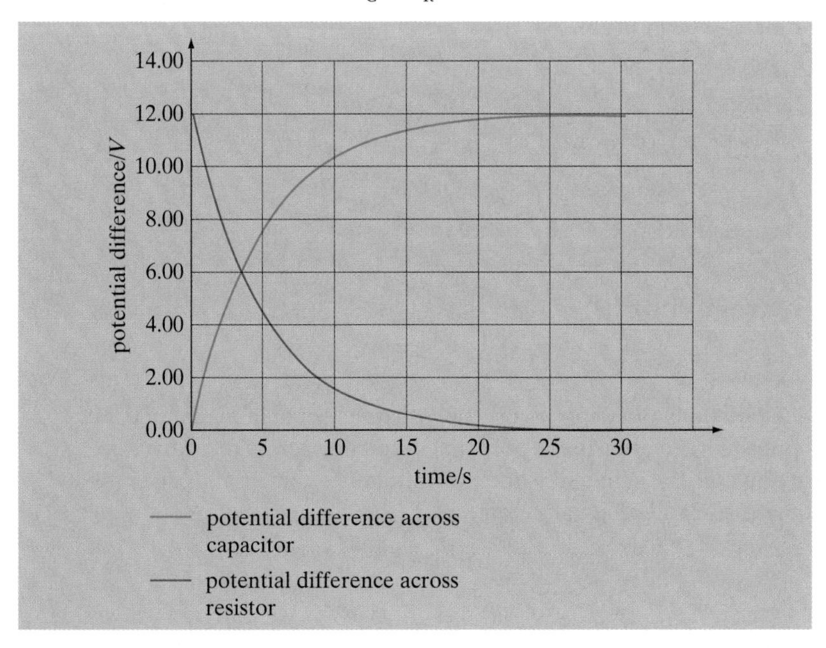

Fig 62 Potential difference versus time in the capacitor-charging circuit of Fig 61

Discharging a capacitor

When a charged capacitor is placed in a circuit to be discharged, the initial discharging current depends on the potential across the capacitor, V, and the resistance of the circuit, R.

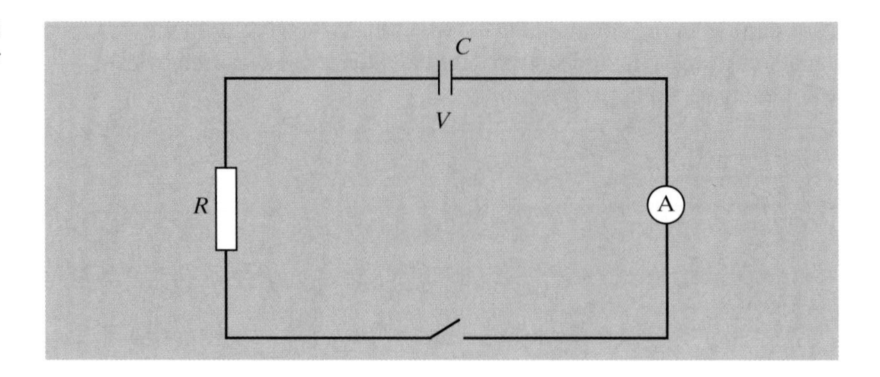

Fig 63
Circuit for discharging a capacitor

The initial current that flows is given by $I_0 = V/R$, and since $V = Q/C$, this can be written $I_0 = Q/CR$. As the charge flows off the capacitor, the voltage drops and the current gets less. The current, charge, and the potential difference across the capacitor all decrease exponentially with time.

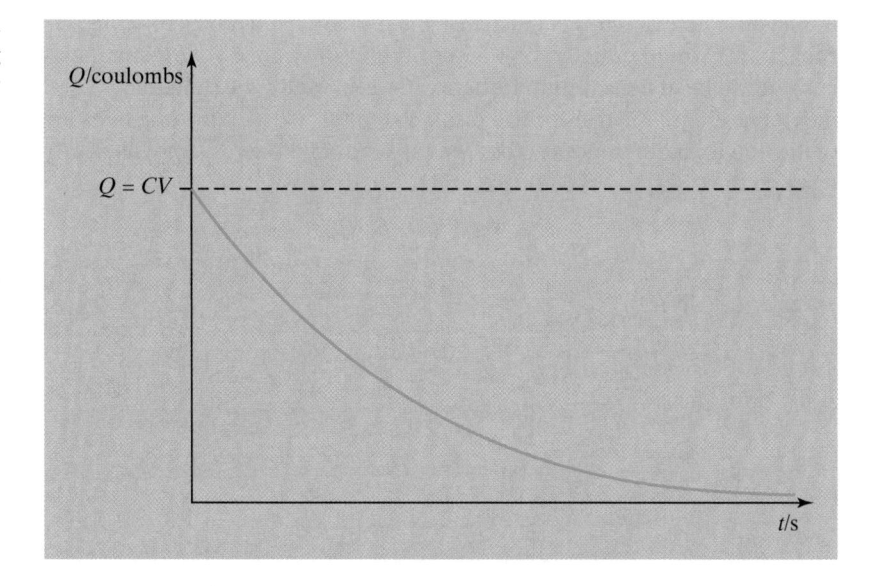

Fig 64
Charge versus time for a discharging capacitor. The graph of V versus time has the same shape

Essential Notes

The current at any time is the gradient of the charge versus time graph. The exponential function has the unique property that the gradient at any point is proportional to the value of the function at that point. The current versus time graph therefore has the same shape as the charge (or potential difference) versus time graph.

The potential difference across the resistor is equal in magnitude, but opposite in polarity, to the potential difference across the capacitor. At any time the sum of the potential differences across the capacitor and the resistor will be equal to zero.

$$V_C + V_R = 0$$

Example

A voltage sensor and a datalogger were used to measure the potential difference across a capacitor as it is discharged through a resistor.

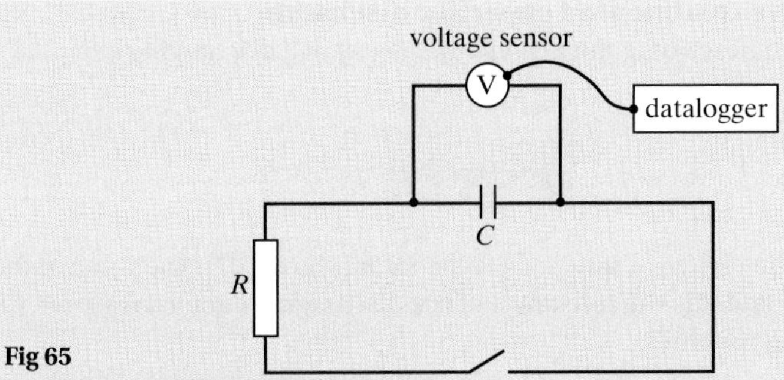

Fig 65

The following results were taken.

Time/s	0	1	2	3	4	5	6
Potential difference/V	10.00	8.19	6.70	5.49	4.49	3.68	3.01
Time/s	7	8	9	10	11	12	
Potential difference/V	2.47	2.02	1.65	1.35	1.11	0.91	

(a) Plot a suitable graph to find the time constant of the circuit.

(b) If the resistor had a value of 100 kΩ, calculate the value of the capacitor.

(c) The resistor and the capacitor are changed to values of 1 kΩ and 5 μF respectively. What sampling rate would you set the datalogger to record at? What would be a reasonable interval over which to record values?

Answers

(a)

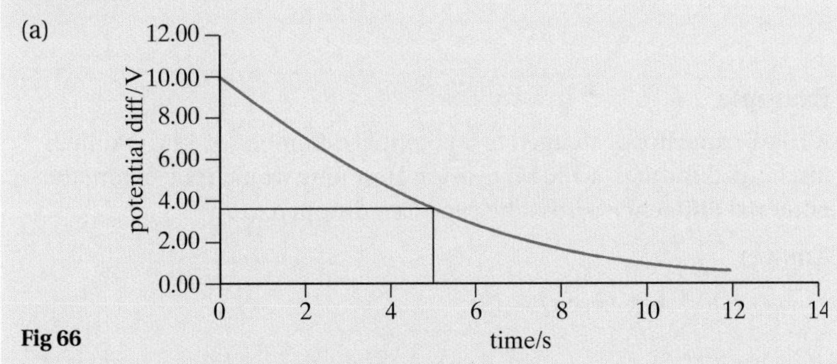

Fig 66

The voltage will drop to 1/e of its original value after 1 time constant. $V = 10/e = 3.687$ V. Reading this value from the graph gives a time constant of 5.0 seconds.

(b) Time constant = $CR = 5.0$ s. So $C = 5.0/100\,000 = 5 \times 10^{-5}$ F = 50 μF.

(c) The new time constant is $1 \times 10^{3}\,\Omega \times 5\,\mu\text{F} = 5 \times 10^{-3}\,\text{s} = 5\,\text{ms}$.

The sampling rate needs to be significantly faster than this, say a minimum of once every 0.5 ms, which is a rate of 2000 Hz. After five time constants the value will have dropped to less than 0.67% of the original voltage. So a logging time of 25 ms should be sufficient.

Quantitative treatment of capacitor discharge

The equation describing the exponential decay of a discharging capacitor is

$$Q = Q_0 e^{-t/CR}$$

where Q is the charge at time t, Q_0 is the initial charge, C is the value of the capacitance and R is the resistance of the discharging circuit. When $t = CR$, this equation becomes

$$Q = Q_0 e^{-CR/CR}$$

$$\frac{Q}{Q_0} = e^{-1} = 0.37 \quad \text{(to 2 sf)}$$

We use logarithms to the base e, sometimes called natural logarithms, to solve problems where we need to find the time taken to drop to a certain charge (or potential).

$$Q = Q_0 e^{-t/CR}$$

Taking natural logs,

$$\ln\left(\frac{Q}{Q_0}\right) = -\frac{t}{CR}$$

So

$$t = -CR \ln\left(\frac{Q}{Q_0}\right)$$

Essential Notes

Don't worry about the negative sign in front of the time. The log of a fraction is always negative so the negative signs will cancel, giving a positive time.

Example

A 10 μF capacitor is charged to a potential difference of 12 V and then discharged through a 100 kΩ resistor. How long would it be before the potential difference across the capacitor dropped to 1 V?

Answer

$Q = Q_0 e^{-t/CR}$, but since $Q = CV$,

$V = V_0 e^{-t/CR}$

Putting in the potential difference values,

$1 = 12 \times e^{-t/CR}$

Since $CR = 10 \times 10^{-6}\,\text{F} \times 100 \times 10^{3}\,\Omega = 1\,\text{s}$,

$1 = 12 \times e^{-t}$ and so $e^{t} = 12$

Taking natural logs,

$t = \ln 12$

giving $t = 2.48\,\text{s}$.

3.4.5 Magnetic fields

Magnetic flux density

A magnetic field can be created by a permanent magnet or by an electric current. A magnetic field is a region where another permanent magnet experiences a force. The field lines around a magnet show the direction of the force on a north pole.

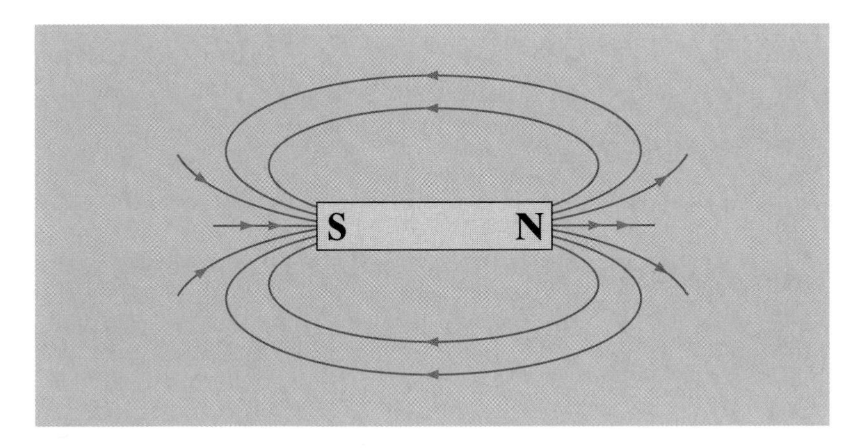

Fig 67
The field lines around a bar magnet

Magnetic fields affect moving charges. Since a flow of current in a wire is due to a stream of moving charges, any current-carrying conductor in a magnetic field experiences a force (Fig 68). This effect, known as the **motor effect**, arises due to the interaction of the magnetic field around the moving charges with the external field.

Essential Notes

Magnetic fields have no effect on stationary charges.

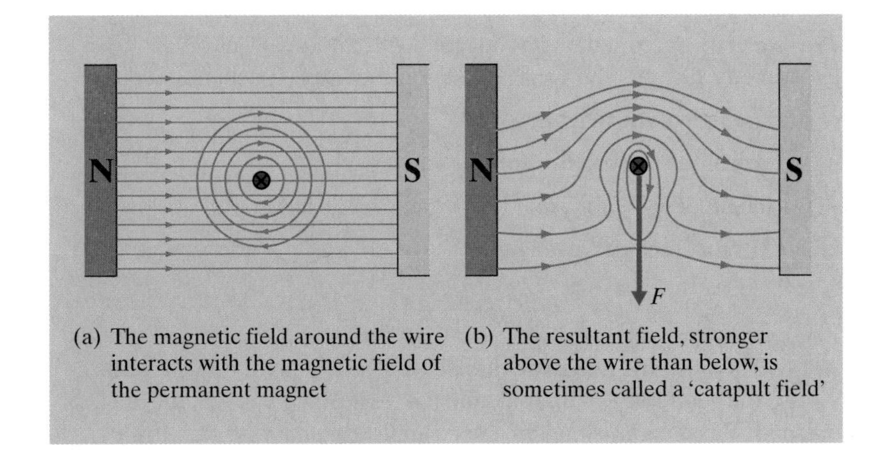

(a) The magnetic field around the wire interacts with the magnetic field of the permanent magnet

(b) The resultant field, stronger above the wire than below, is sometimes called a 'catapult field'

Fig 68
Catapult field around a current-carrying wire, giving rise to a force

Essential Notes

A cross, ×, is used to show a current, or a magnetic field line, pointing away from you into the plane of the paper. A dot is used to show a current, or a magnetic field line, pointing towards you out of the plane of the paper.

The direction of the force is at right angles to both the direction of the current and the direction of the field. Fleming's left-hand rule (Fig 69) is a way of remembering the relative directions of the magnetic field, the electric current and the resulting force.

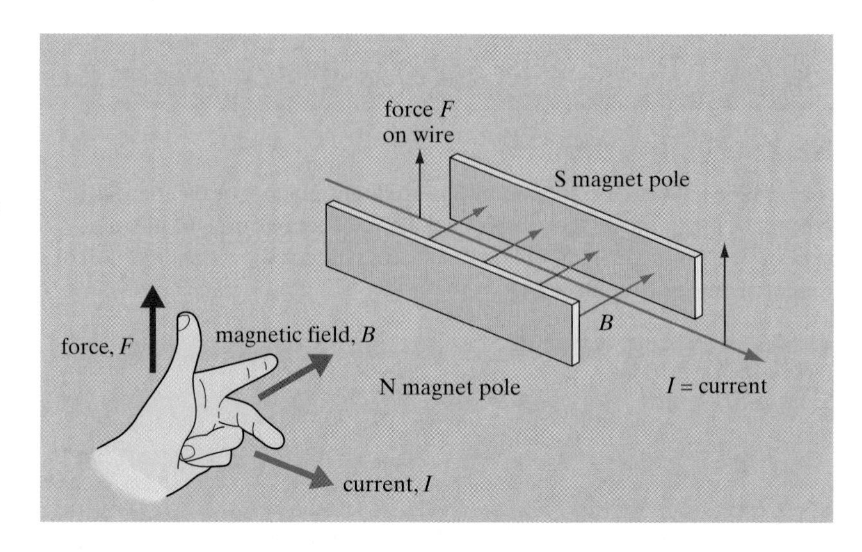

Fig 69
Fleming's left-hand rule gives the direction of the force

The first finger must point in the direction of the field, from north towards south. The second finger must point in the direction of conventional current, from + to −. The thumb will then indicate the direction of the force on the wire

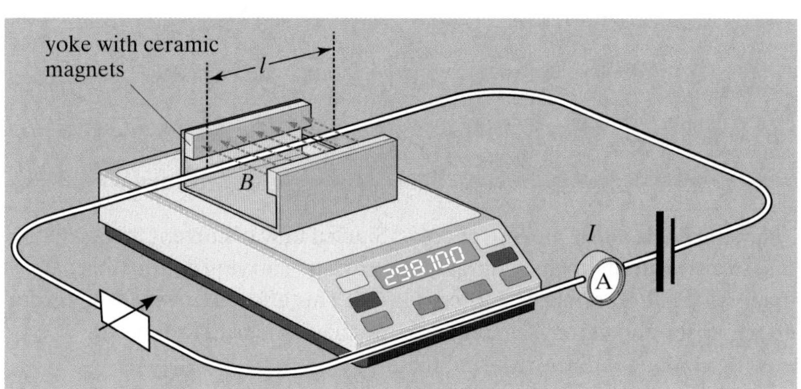

Fig 70
Apparatus to investigate the size of force on a current-carrying wire

The size of the force can be investigated with the apparatus in Fig 70. Experiments like this show that the size of the force, F, dependent on:

- the strength of the magnetic field;
- the size of the electric current, I;
- the length of the wire, l, that is in the magnetic field.

In SI units the force F is given by

$$F = BIl$$

Where B is called the **magnetic flux density** which has SI units of **tesla**,T. We use the equation $F = BIl$ to define the tesla. Since $B = F/Il$, the magnetic flux density is the force per unit length on a wire carrying a current of one ampere.

Essential Notes

If the current and the wire are not at right angles, the force becomes $F = BIl \sin \theta$ where θ is the angle between the current and the field.

Definition

One tesla is the magnetic flux density that causes a force of 1 N to be exerted on every 1 m length of a wire carrying 1 A of current in a direction that is perpendicular to the field.

The tesla is quite a large unit; a typical permanent magnet would have a flux density close to the poles of around 100 mT, whilst the strongest electromagnets in the world create maximum flux densities of only a few tens of tesla.

Moving charges in a magnetic field

The force on a current-carrying wire in a magnetic field is given by $F = BIl$. This equation can be adapted to apply to a charge Q that is moving in a magnetic field at a velocity v.

$$I = \frac{Q}{t}, \text{ so } F = B\left(\frac{Q}{t}\right)l$$

This can be written as

$$F = BQ\left(\frac{l}{t}\right)$$

and since $\frac{l}{t}$ = velocity, v, the equation becomes

$$F = BQv$$

Definition

The force on a charged particle moving through a magnetic field is F = BQv, where Q is the charge of the particle and v is its velocity perpendicular to the magnetic field.

Because the force is always at right angles to the particle's velocity, a charged particle moving in a magnetic field has a circular path. The magnitude of the particle's velocity does not change, but its direction is constantly changing.

The magnetic force, BQv, causes a centripetal acceleration. For a particle of mass m,

$$BQv = \frac{mv^2}{r}$$

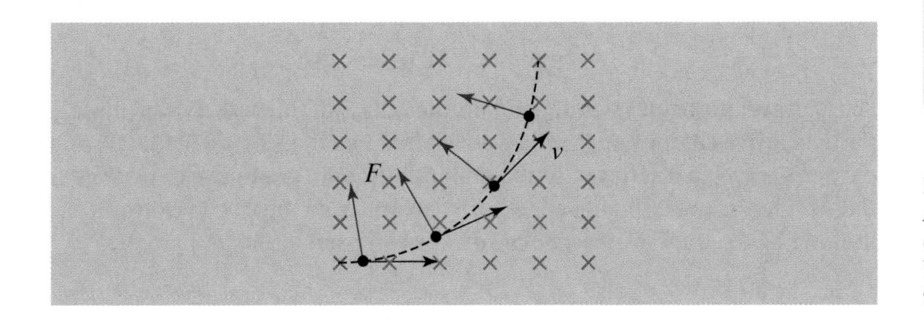

Examiners' Notes

Use Fleming's left-hand rule (page 62) to find the direction of the force. The second finger (current) points in the direction that a positive charge is moving. Remember to reverse the direction of your second finger if it is a negative particle, such as an electron.

Fig 71
A moving charged particle follows a circular track in a magnetic field because the force on it is always at 90° to its velocity

The radius of the path is therefore given by

$$r = \frac{mv}{BQ}$$

Magnetic fields are used to deflect charged particles in particle accelerators like the cyclotron. In a cyclotron a source of charged particles is placed in the centre of two 'D'-shaped regions. A magnetic field curves the particles' path into a semicircle in each D.

Fig 72
A cyclotron

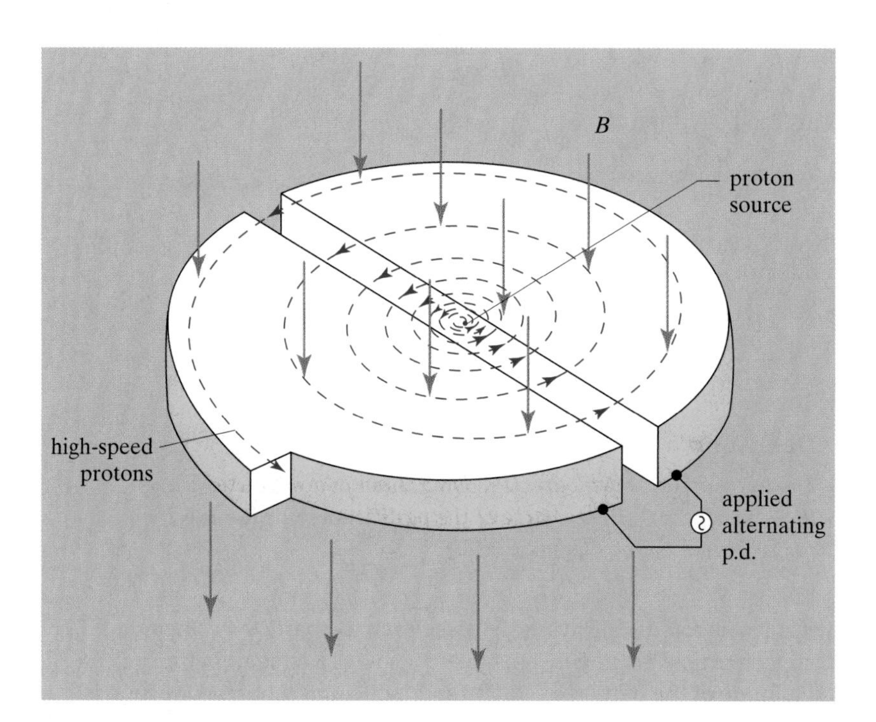

An alternating potential difference accelerates the particles as they pass across the gap between the Ds. For the cyclotron to work, the time taken for a particle to travel one half-circle must match the time taken for the polarity of the a.c. to reverse, so that the particles arrive at the gap at just the right time to be accelerated. The time taken, t, for a particle to travel a whole circle is

$$t = \frac{2\pi r}{v} = 2\pi r \times \frac{m}{BQr} = \frac{2\pi m}{BQ}$$

So the travel time does not depend on the radius of the path. Every time a particle arrives at the gap it will be accelerated by the electric field. The particles travel in a series of semicircular paths with gradually increasing radius. They move at higher speeds on each turn, so that the cyclotron's limiting factor is its size: bigger cyclotrons can produce faster particles.

Essential Notes

Increasing the size of a cyclotron only works up to a certain limit. As the particles approach the speed of light their mass increases and the orbital time changes. The relativistic increase in mass is covered in optional unit 5D.

Magnetic flux and flux linkage

The magnetic flux density can be thought of as the number of flux lines that pass through a given area. In a strong field the field lines are close together and the flux density is high.

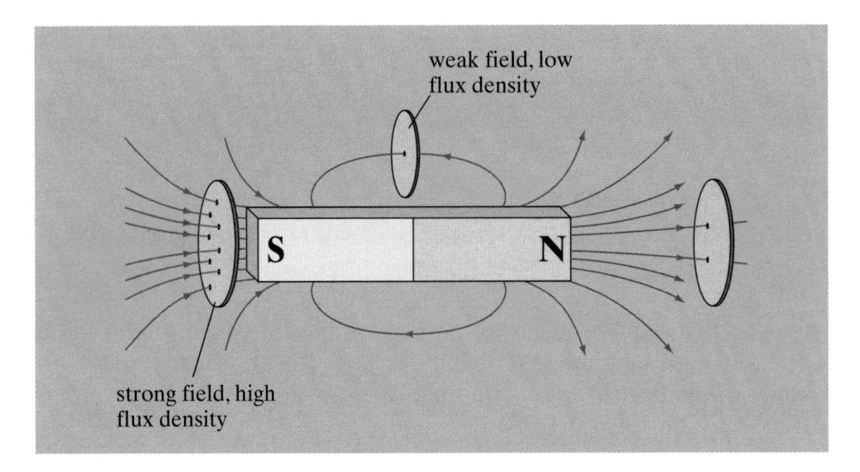

Fig 73
Flux density and field lines

Provided that the area is at right angles to the field lines, the total **flux**, Φ, that passes through a given area, A, is given by

$$\text{total flux} = \text{flux density} \times \text{area}$$

or

$$\Phi = BA$$

Flux is measured in units of **webers**, Wb.

Essential Notes

Since flux density is the flux per unit area, one tesla is equal to 1 weber per square metre.

Sometimes, especially when we arc considering electromagnetic induction (see below), it is useful to calculate the total flux through a coil of wire. If the coil has several turns of wire, the flux through the whole coil is the sum of the flux through each individual turn. This is referred to as the **flux linkage** through the coil. For a coil of N turns the flux linkage is $N\Phi$.

The amount of flux linkage through a coil depends on the orientation of the coil with respect to the magnetic field. If θ is the angle between the normal to the plane of the coil and the magnetic field (Fig 74), the total flux Φ through the rectangular coil is $BA\cos\theta$. If there are N turns on the coil, the total flux linkage is

$$N\Phi = BAN\cos\theta$$

Fig 74
Flux linkage is $BAN\cos\theta$

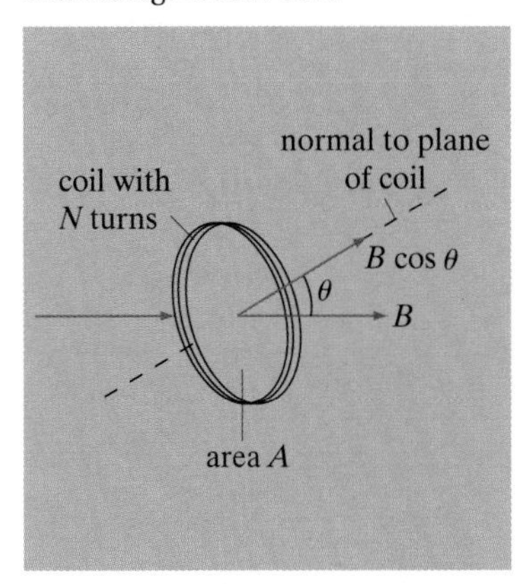

Electromagnetic induction

When a bar magnet is moved relative to a coil of wire connected in a circuit, an electric current is made to flow in the coil. The potential difference that causes this current is referred to as an **induced e.m.f.** The direction of the induced e.m.f. depends on the direction of motion.

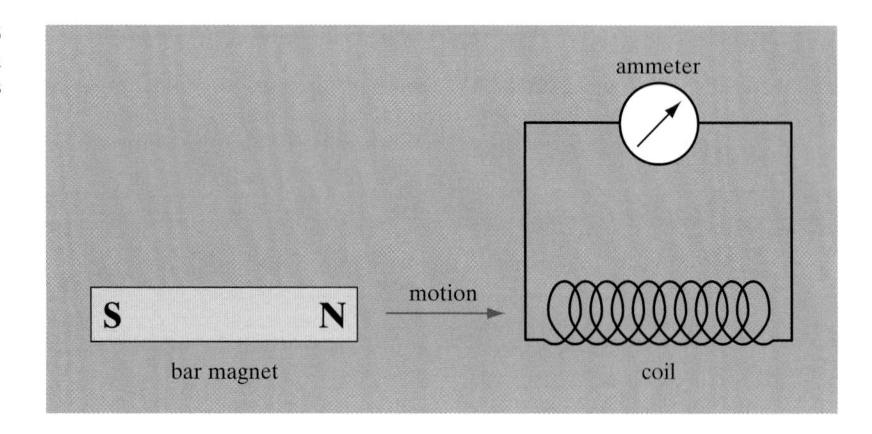

Fig 75
Simple electromagnetic induction apparatus

As the north pole of the magnet is pushed towards the coil, the induced e.m.f. will cause a current to flow in the coil (Fig 75). This current will create a magnetic field around the coil. There are two possibilities:

1 The end of the coil nearest the magnet could become a south magnetic pole. This would attract the bar magnet, thereby increasing the rate at which flux changes through the coil. This would increase the induced e.m.f. and current and make the magnet stronger, which would attract the magnet with a greater force. This sequence of events would continue and would increase the kinetic energy of the magnet at the same time as increasing the electrical energy generated. This would contravene the conservation of energy and so it does not happen.

2 The end of the coil nearest the magnet could become a north pole. This would repel the bar magnet. Indeed the quicker you tried to push the magnet into the coil, the larger the induced e.m.f., and the larger the current and the greater the force opposing the motion.

Definition

Lenz's law *states that the direction of the induced e.m.f. always acts so as to oppose the change that is causing the e.m.f.*

Faraday's law

We can use the same apparatus (Fig 75) to investigate the factors that affect the magnitude of the induced e.m.f.

The size of the e.m.f. that is induced in the coil depends on:

- the relative speed of the coil and the magnet: the faster the movement, the larger the e.m.f.;
- the number of turns on the coil;
- the area of the coil;
- the strength of the magnetic field.

Faraday combined these effects into a single statement.

Definition

***Faraday's law** of electromagnetic induction states that the magnitude of the induced e.m.f., E, is proportional to the rate of change of magnetic flux linkage through the coil.*

For a coil of N turns, this can be written as

$$E = \frac{N\Delta\Phi}{\Delta t}$$

Example

A small circular coil of radius 2 cm is placed close to a bar magnet so that its plane is perpendicular to the field lines. The magnetic flux density is 40 mT. The coil, which has 20 turns, is turned so that it is edge-on to the field. If this movement takes 0.1 s, calculate the induced e.m.f.

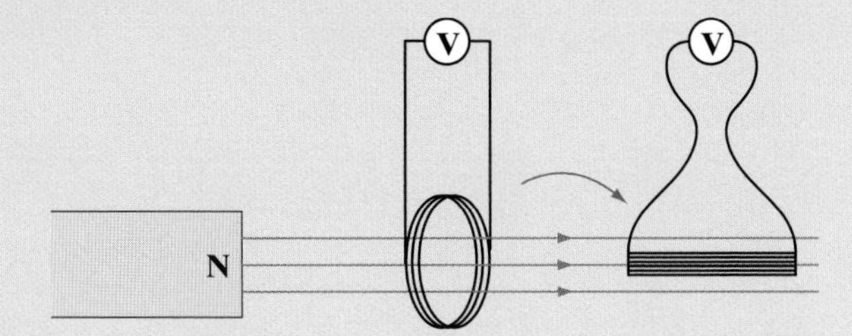

Fig 76

Answer

Initially, the total flux linkage through the coil is:

$N\Phi = BAN = 40 \times 10^{-3} \times \pi(0.02)^2 \times 20 = 1.01 \times 10^{-3}$ Wb turn

When the coil is turned the flux linkage drops to zero, so

$N\,\Delta\Phi = 1.01 \times 10^{-3}$ Wb turn

So

$$E = N\,\Delta\Phi/\Delta t = \frac{1.01 \times 10^{-3}}{0.1} = 0.01\text{ V}$$

Faraday's law applies more generally to any situation where there is relative motion between a conductor and a magnetic field. Another way of thinking about the law is to say that the induced e.m.f. is proportional to the rate at which field lines are cut through.

An aircraft flying horizontally cuts through the vertical component of the Earth's magnetic field. An e.m.f. is generated between the tips of its wings. To calculate the e.m.f. we need to calculate the flux cut through per second.

Area travelled through in 1 second A = length of wings × velocity = $l \times v$

Flux cut through in 1 second gives the induced e.m.f.:

$$E = BA = Blv$$

The vertical component of the Earth's magnetic field is 4×10^{-5} T so if the wingspan of the aircraft is 40 m and the aircraft's speed is 130 m s^{-1}, the induced e.m.f. will be

$$E = 4 \times 10^{-5} \times 130 \times 40 = 0.21\ \text{V}$$

The e.m.f. induced in a rotating coil

A simple generator can be constructed from a coil rotating in a magnetic field, Fig 77.

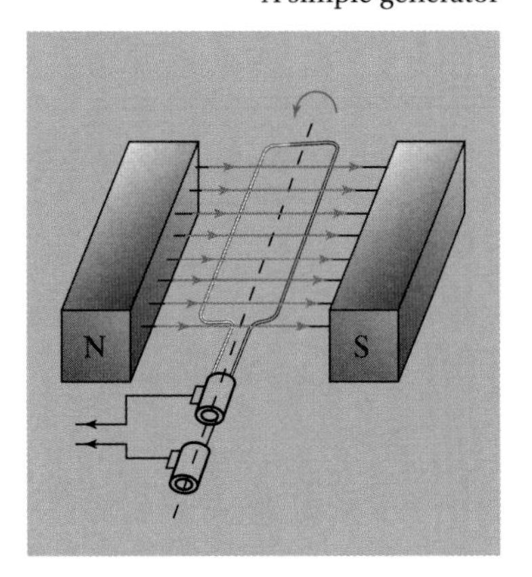

Fig 77
A simple generator

The induced e.m.f. in the coil is equal to the rate of change of magnetic flux linkage, according to Faraday's law. Therefore, whenever the coil is turning and cutting through magnetic field lines, an e.m.f. will be induced. The greater the rate of cutting field lines, the greater the e.m.f.

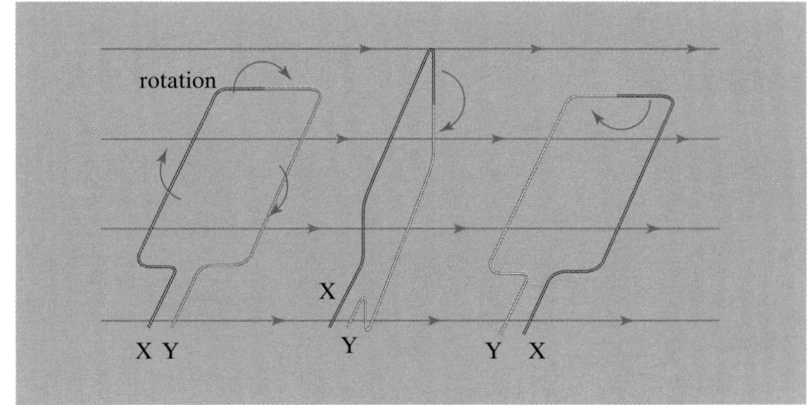

Fig 78
Changing flux linkage through the rotating coil

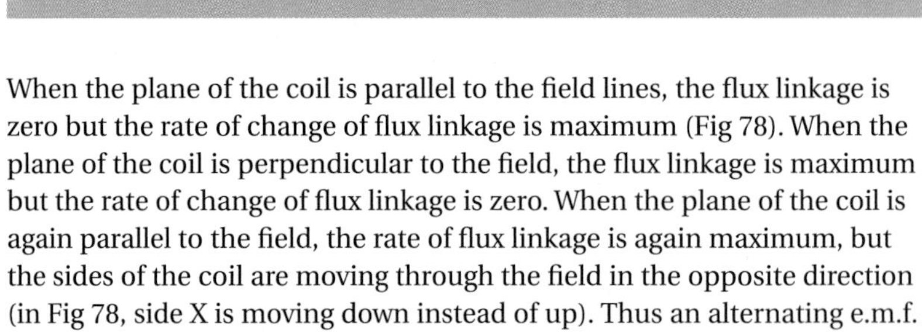

When the plane of the coil is parallel to the field lines, the flux linkage is zero but the rate of change of flux linkage is maximum (Fig 78). When the plane of the coil is perpendicular to the field, the flux linkage is maximum but the rate of change of flux linkage is zero. When the plane of the coil is again parallel to the field, the rate of flux linkage is again maximum, but the sides of the coil are moving through the field in the opposite direction (in Fig 78, side X is moving down instead of up). Thus an alternating e.m.f. is induced, as shown in Fig 79.

The flux linkage through the coil is $BAN \cos\theta$ (see page 65). The e.m.f. induced across the coil is given by the rate of change of flux linkage = $\Delta(BAN\cos\theta)/\Delta t$, which is the gradient of graph 1 in Fig 79. Graph 2 shows how the gradient of graph 1 changes with time. This is the induced e.m.f; it varies sinusoidally with time.

If the coil rotates with angular frequency $(\Delta\theta/\Delta t) = \omega$ rad s^{-1}, the induced e.m.f. at time t is given by

$$E = BAN\omega \sin \omega t$$

This varies between 0 and $BAN\omega$.

Essential Notes

We can derive the expression for the induced e.m.f. using calculus. Differentiating the expression for flux linkage $BAN \cos\theta$ with respect to time to obtain the rate of change of flux linkage, with $\theta = \omega t$, gives $-BAN\omega \sin \omega t$ for the induced e.m.f. The negative sign shows that the e.m.f. is in a direction that opposes the rotation, in accordance with Lenz's law.

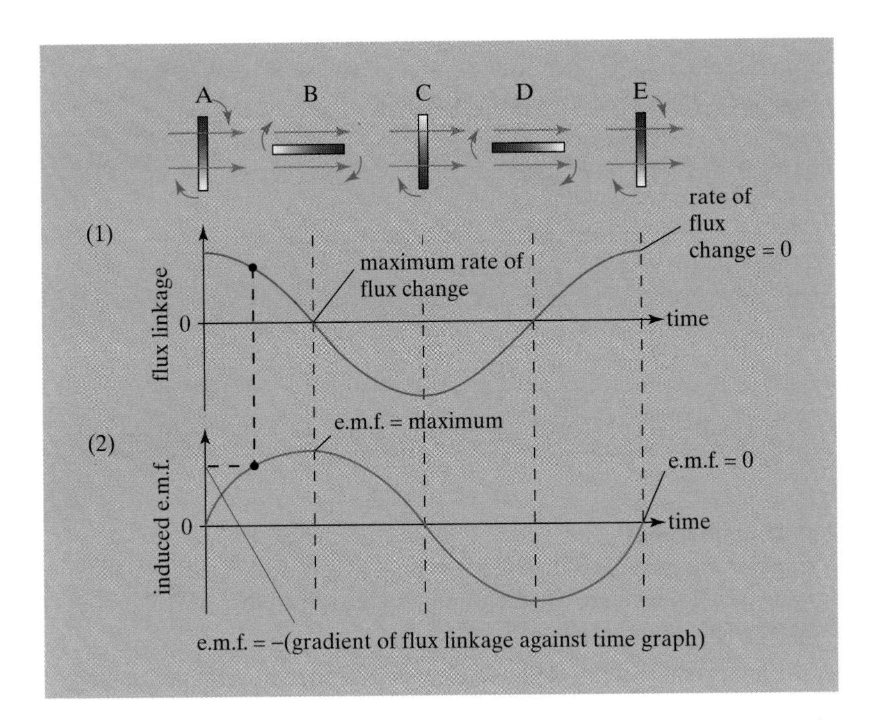

Fig 79
Induction of an alternating e.m.f.

A: The coil is face-on to the field and flux linkage is maximum
B: The coil is edge-on to the field and flux linkage is zero
C: The coil is face-on to the field but moving in the opposite direction to A
D: The coil is edge-on to the field and flux linkage is zero
E: The coil is face-on to the field and flux linkage is maximum

Example

A bicycle dynamo is made from a circular coil of 100 turns of wire which rotates between two magnets where the flux density is 40 mT. The radius of the coil is 1.5 cm.

(a) If the coil rotates at 1500 rpm, calculate the peak voltage produced.

(b) The e.m.f. generated by the dynamo is shown on an oscilloscope. Sketch the trace you would expect to see if the bicycle travelled twice as fast.

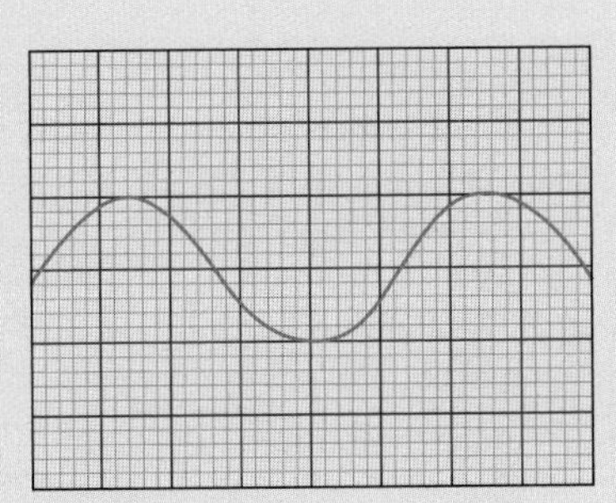

Answer

(a) Peak e.m.f. $= BAN\omega$

$A = \pi r^2 = 3.14 \times (1.5 \times 10^{-2})^2 = 7.07 \times 10^{-4}\,\text{m}^2$

$\omega = 2\pi f = 2\pi \times 1500/60 = 157\,\text{rad s}^{-1}$

Peak e.m.f. $= 40 \times 10^{-3} \times 7.07 \times 10^{-4} \times 100 \times 157 = 0.44\,\text{V}$

Examiners' Notes

From Faraday's law, since the rate at which the flux linkage changes will double, the peak induced e.m.f. will also double. The dynamo will turn twice as fast so the coil will go through twice as many cycles per second (see Fig 79). This will double the frequency of the a.c.

(b) The peak voltage (amplitude) of the signal would double. The frequency of the signal would also double.

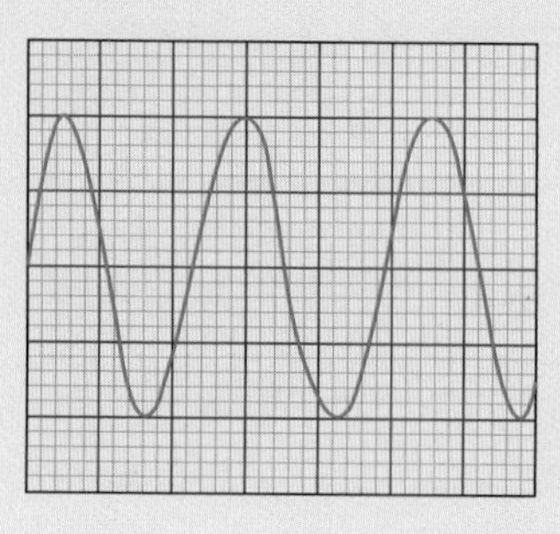

Fig 80
The principle of a transformer

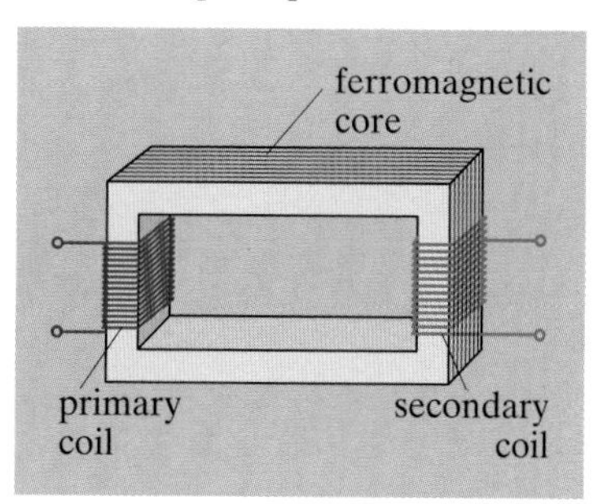

The transformer

A **transformer** is used to change the voltage of an a.c. signal. It is made from two coils which are usually wound on the same piece of ferromagnetic material (such as soft iron).

The primary coil is connected to an a.c. source, so there is an alternating current in the primary coil. The primary coil acts as an electromagnet which varies in strength as the current through it varies. The magnetic flux in the core therefore changes, changing the flux linkage through the secondary coil. This induces an alternating e.m.f. in the secondary coil.

If there are fewer turns on the secondary than on the primary, the output e.m.f. is smaller than the input voltage, and the transformer is said to be a step-down transformer. If there are more turns on the secondary than on the primary, the output voltage is greater than the input voltage, and the transformer is said to be a step-up transformer.

Fig 81
A step-down transformer

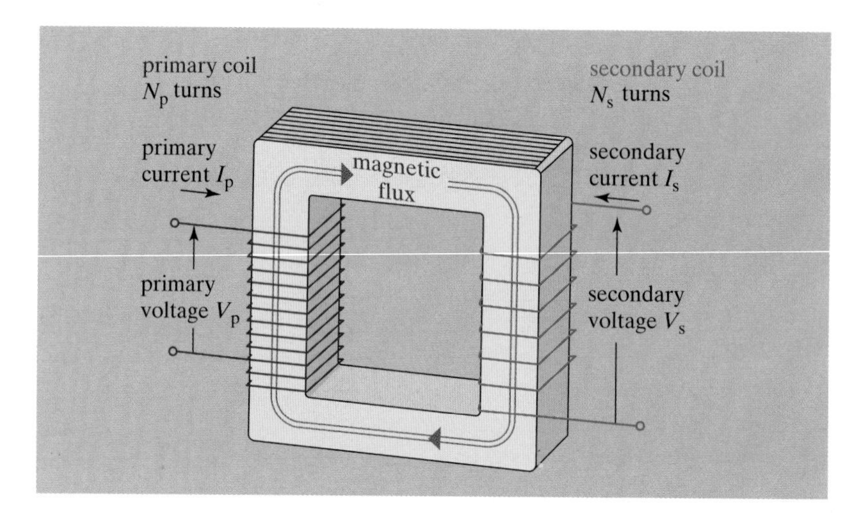

The ratio of the voltages across the primary and secondary coils is equal to the ratio of the number of turns.

Definition

The transformer equation is $\frac{N_s}{N_p} = \frac{V_s}{V_p}$

It may seem that a step-up transformer is generating energy from nowhere. This of course would contravene the principle of conservation of energy. In fact, a step-up transformer increases the output voltage but decreases the output current. If the efficiency of the transformer was 100%, then all of the electrical power on the primary side of the transformer would be transferred to the secondary side. Since power $P = IV$,

$$I_p V_p = I_s V_s \quad \text{for 100\% efficiency}$$

A step-up transformer that increases the voltage by a factor of 100 must therefore lead to a decrease in the current by a factor of 100.

In practice the efficiency is always less than 100%. The efficiency of the transformer is given by

$$\text{efficiency} = \frac{\text{output power}}{\text{input power}} = \frac{I_s V_s}{I_p V_p}$$

Transformers are less than 100% efficient for the following reasons.

- Energy is transferred as heat in the windings. Power loss in the primary and secondary coils is due to the electrical resistance of the coils. The efficiency can be improved by reducing the resistance of the coils – by using thicker copper wire or even by using superconducting coils, as in some experimental transformers.
- Energy is transferred as heat in the ferromagnetic core. The changing flux in the core leads to 'eddy currents' being induced in the core. The size of the eddy current is reduced by using a core that is laminated – sheets of iron are interleaved with sheets of an insulating material.
- The changing magnetic flux leads to fluctuating forces between the primary and secondary coils, and between the metal sheets in the laminated core. These varying forces lead to the buzzing that can sometimes be heard coming from transformers.
- Not all of the flux generated by the primary intercepts the secondary. This does not necessarily lead to energy losses, but if this leakage flux intercepts a conductor, there will be eddy currents. Flux linkage can be improved by using co-axial coils, where the secondary is inside the primary.
- There are also hysteresis losses in the core – as the core is magnetised and demagnetised, some energy is dissipated in heating it. These losses can be reduced by using new materials for the core which are easily magnetised and demagnetised.

Small transformers, such as those used for laptop computers or for charging mobile phones, are often less than 85% efficient. Larger transformers, such as those used in the National Grid, are likely to be better than 98% efficient.

The National Grid

Power stations that generate electricity are often a long distance from where the electricity is needed. A transmission system, the National Grid, is used to transmit the electricity. There are significant energy losses due to the resistance of the cables used. These energy losses are due to heating of the wires and are dependent on the current, *I*, that flows in the wires. The power loss due to resistance, *R*, is given by

$$P_{\text{loss}} = I^2R$$

In order to keep the power loss as low as possible, it is important to transmit the electrical power at low current. However, the power delivered from the power station is $P = IV$. To transmit high power, at low current, a very high potential is needed. Step-up transformers are therefore used at power stations to increase the voltage to a very high value, 400 kV, for transmission over long distances. Step-down transformers are used near to towns and industrial sites to reduce the potential to safer values.

Fig 82
Electrical energy generated at 25 kV is stepped up for long-distance transmission via the Supergrid to 400 kV, then down to 132 kV for local distribution networks. This is stepped down further for use in factories, and down to the standard domestic voltage of 230 V for home use

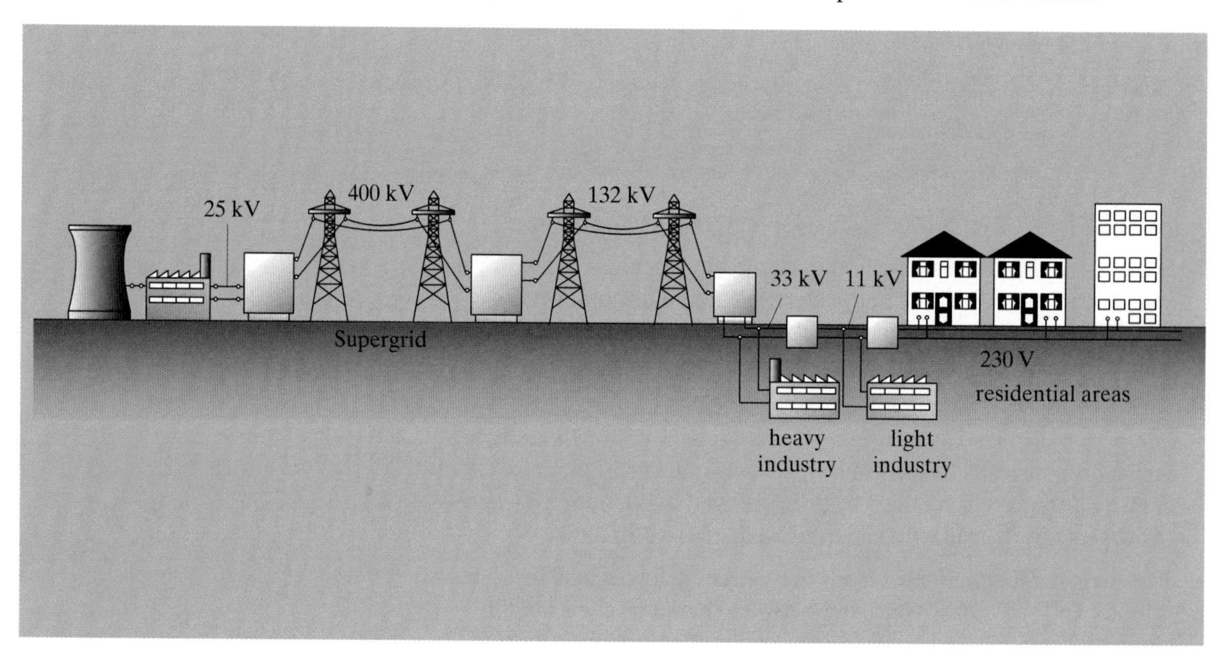

Example

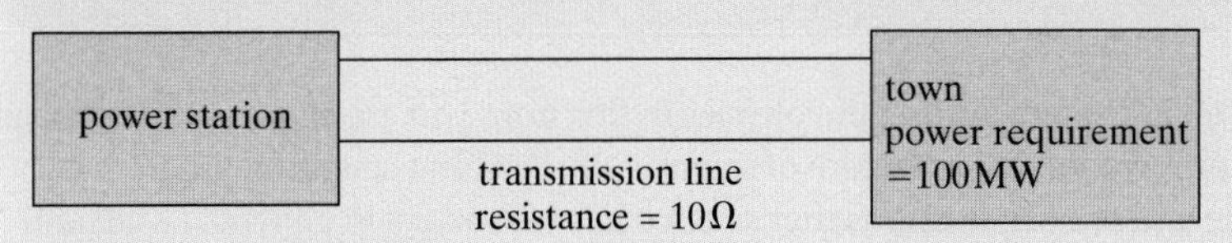

Fig 83

A power station is connected to a town by cables of resistance 10 Ω.

1 Calculate the power losses in the cables if:

(a) The power is supplied to the town at 20 kV.

(b) The power is supplied to the town at 400 kV.

2 Compare the efficiency of transmission at low voltage and at high voltage.

Answers

1 (a) The power required at the town is 100 MW. The current is found from $P = IV$, giving

$$I = \frac{P}{V} = \frac{100 \times 10^6}{20 \times 10^3} = 5000 \text{ A}$$

Power losses in the cables = $I^2R = 5000^2 \times 10 = 250$ MW

(b) Similarly,

$$I = \frac{P}{V} = \frac{100 \times 10^6}{400 \times 10^3} = 250 \text{ A}$$

Power losses in the cables = $I^2R = 250^2 \times 10 = 625$ kW

2 At low voltage the power station has to provide a total power of $100 + 250 = 350$ MW, so the efficiency is

$$\frac{\text{useful power}}{\text{total power supplied}} = \frac{100}{350} = 29\%$$

At high voltage the power station has to provide a total power of $100 + 0.625 = 100.625$ MW, so the efficiency is

$$\frac{\text{useful power}}{\text{total power supplied}} = \frac{100}{100.625} = 99.4\%$$

How Science Works

As well as understanding the physics in this unit, you are expected to develop an appreciation of the nature of science, the way that scientific progress is made and the implications that science has for society in general. GCSE and A-level science syllabuses refer to these areas as '*How Science Works*'.

The *How Science Works* element of your course, which also includes important ideas about experimental physics, may be assessed in the written examination papers as well as in the internally assessed unit, the Investigative Skills Assessment or ISA. The concepts included in *How Science Works* may be divided into several areas.

Theories and models

Physicists use theories and models to attempt to explain their observations of the universe around us. These theories or models of the real world can then be tested against experimental results. Scientific progress is made when experimental evidence is found that supports a new theory or model.

You are expected to be aware of historical examples of how scientific theories and models have developed and how this has changed our knowledge and understanding of the physical world.

An example of a scientific model is the kinetic theory of gases. An ideal gas is modelled by making several simplifying assumptions, such as that the molecules of the gas occupy zero volume. This is close to the truth, since the total volume occupied by the molecules is very much less than the volume of the gas itself. We also assume that there are no forces between the molecules except during collisions. Again this is almost true since the intermolecular forces are short range and the gas molecules are, on average, a relatively long way apart. When we apply Newton's laws of motion to the molecules of this ideal gas we are able to derive expressions for temperature and pressure which are in close agreement with the experimentally derived gas laws, such as Boyle's law.

Experimental results can of course disprove a model that was previously accepted. Rutherford's scattering experiment showed that the 'plum pudding' model of an atom could not be correct. Rutherford's results led to the development of a new model, that of the nuclear atom.

You should know the meaning of the terms 'hypothesis' and 'prediction'.

- A **hypothesis** is a tentative idea or theory, or explanation of an observation.
- A **prediction** from a hypothesis or theory is a forecast that can be tested by experiment.

If a reliable experiment does not support a hypothesis, then the hypothesis is likely to be abandoned or modified. Hypotheses are not usually widely accepted until the experimental results have been repeated by a number of independent scientists. It may take many experimental tests until a set of hypotheses become accepted as a scientific theory (see the figure left).

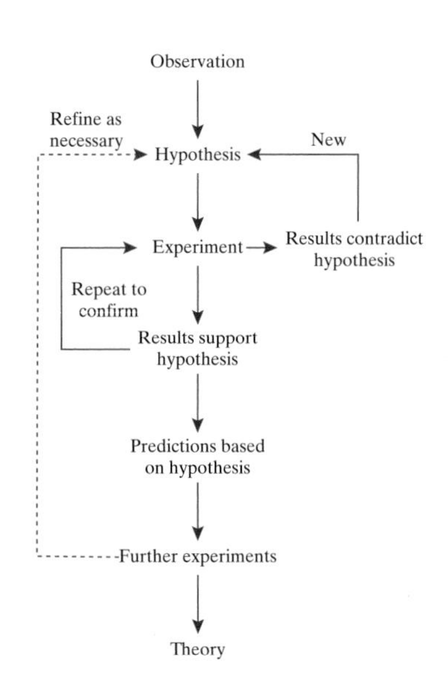

The stages of scientific research

Even then, a scientific theory is always capable of being later refuted if compelling experimental evidence suggests that a new explanation is necessary.

Experimental techniques

You are expected to develop the skills of experimental planning, observation, recording and analysis. These skills will mainly be assessed in the practical coursework, the ISA, or by the externally-marked practical assignment, the EMPA. This section contains some general advice for carrying out experimental work in physics.

When you plan an experiment you need to be able to identify the dependent, independent, and control variables that are involved.

- The **independent variable** is the physical quantity that you deliberately change.
- The **dependent variable** changes as a result of this.

For example, if you are asked to investigate how the length of a piece of metal wire affects its electrical resistance, the length is the independent variable, and the electrical resistance is the dependent variable. Any other variables that may have an impact on the outcome need to be controlled so that the conclusions of the experiment are clear. These are known as the **control variables**. In the example of the wire, two of the control variables are the cross-sectional area of the wire and its temperature.

You will need to select appropriate apparatus, including measuring instruments of a suitable precision and accuracy. These two terms are often confused.

- **Accuracy** refers to how close the reading is to the accepted value.
- **Precision** refers to the number of significant figures that the measurement is made to.

For example, an electronic balance that gives an answer to 0.01 g, e.g. 3.24 g, is capable of more precise measurements than a balance that measures in grams only, e.g. 3 g. If several readings of the same measurement are closely grouped together, the readings are said to be **precise**. If the readings agree with a known mass, they are said to be **accurate**. The analogy of rifle shots at a target may be used to differentiate between accuracy and precision (see the figure right).

A accurate

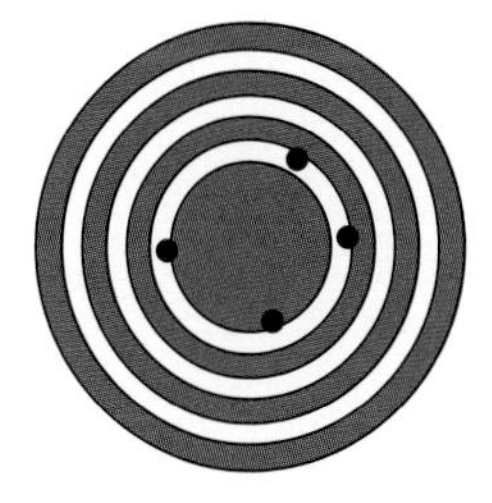

B precise

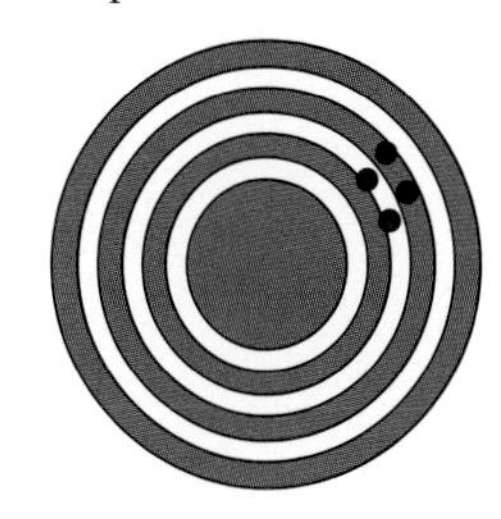

Figure A shows accurate shooting, since the bullets (or readings) are close to the centre of the target. But the shooting is not precise, since the bullets are widely scattered. Figure B shows precise shooting, since the bullets (readings) are closely grouped, but not accurate, since the bullets (readings) are not close to the centre of the target (accepted value).

When choosing your apparatus you should be aware that ICT can be used to assist with the collection and analysis of experimental data. This may mean using a suitable sensor, attached to a data logger, to take the readings and then a spreadsheet to help to analyse them. For example, suppose that you wanted to investigate the current surge that passes through the filament of an incandescent light bulb when it is first turned on. The light bulb reaches its operating temperature very quickly, so the readings need to be taken rapidly. A current sensor attached to a data logger could take the readings at the required rate. Conversely, sometimes the readings need to be taken over a long time period, for example if you wanted to investigate the cooling of a house overnight. A temperature sensor and a data logger would make the job somewhat less tedious!

Planning an experiment also means being aware of any risks to health and safety, and taking any necessary precautions. For example, you need to wear protective goggles when stretching a metal wire.

You also need to plan to reduce experimental errors. There are two types of error: systematic and random.

- **Systematic errors** cannot be reduced by repeating the measurement; for example, using an electronic balance which is not zeroed would lead to a systematic error.
- **Random errors** occur when taking readings, such as the timing errors when measuring the period of a pendulum. These can be reduced by repeating the readings and finding the mean, since the errors are random and may fluctuate above and below an average reading.

It is important to identify the percentage uncertainty associated with each reading. For example, if a length is measured using a ruler with millimetre divisions the reading in centimetres may be given to a precision of ± 0.1 cm. A reading of 25.2 ± 0.1 cm has a percentage uncertainty of $({}^{0.1}/_{25.2}) \times 100\% = 0.4\%$.

The percentage uncertainty can be reduced by using a more precise measuring device, such as a micrometer or vernier callipers instead of a ruler. The percentage uncertainty can also be reduced by increasing the size of the quantity to be measured. For example, to measure the thickness of a sheet of paper, you could measure the thickness of 100 sheets of paper, and then divide by 100 to find the value for a single sheet. When timing the period of a pendulum it is good practice to time a number of oscillations, say 10, and then divide by 10.

When you tabulate your readings you should ensure that the columns are headed with the quantity *and* the unit that it is measured in. It is good practice to also include the uncertainty associated with that reading in the heading, as in the table here.

Length/m ±0.001 m	Time/s ±0.1 s
0.023	2.3
0.031	3.1
0.04*	4.0

* see below

You should always quote figures in your results table to the appropriate degree of precision, and be consistent. The length reading in the last row of the table isn't correct – it should read 0.040.

The most significant uncertainty in the readings taken in an experiment may determine whether you can draw a reliable conclusion or not.

You can also test reliability by repeating the experiment a number of times and comparing your results. In practice scientists share their results through publication. Reliability is tested by other scientists who try to replicate the work, or check the results using a different experimental method.

When you plot your results on a graph, you should choose a scale so that the range of your points covers at least half of the graph paper, in both the x- and y-directions. Choose scales that have divisions that are easy to interpret accurately, i.e. divisions in multiples of 2, 5 or 10 rather than 3 or 7.

You must label each axis of the graph with the quantity and the unit; conventionally this is done using a solidus, for example length/m. The best-fit line or curve should be drawn so as to minimise the total distance of points from the line. In practice this will mean that some points fall above the line and some below (see the graph here). Any results that fall outside the expected range of values (anomalous values) can be ignored when choosing the best line, but you should identify these anomalous readings on your graph. You should also repeat these readings whenever possible and try to explain why they do not follow the trend. Finally, when measuring the gradient of a line, choose a large section of the graph so as to reduce errors.

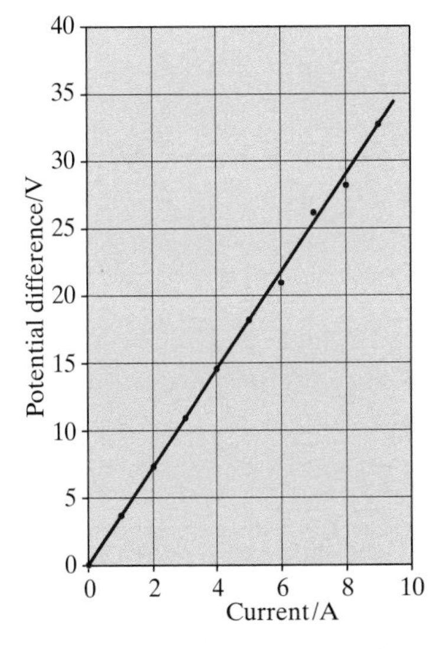

Graph to show how the potential difference across a wire affects the current through it

Applications and implications of science

Scientific advances have greatly improved the quality of life for the majority of people, and developments in technology, medicine and materials continue to further these improvements at an increasing rate. However, technologies themselves pose significant risks that have to be balanced against the benefits. For example, the way in which electrical energy is generated poses many questions. Nuclear power is capable of generating large amounts of energy and does not emit carbon dioxide, but it does produce radioactive waste which needs to be stored safely for thousands of years.

Scientific findings and technologies enable advances to be made that have potential benefit for humans; however, the scientific evidence available to policy makers may be incomplete. Political decision-makers are influenced by many things, including prior beliefs, vested interests, public opinion and the media, as well as by expert scientific evidence.

Guidance on logarithms and logarithmic graphs

In your A2 examinations, you may be asked to construct or interpret graphs with a logarithmic scale. If you are studying A-level Mathematics you may have met logarithms already. This section tells you what you need to know for A-level Physics.

There are two main reasons for using logarithmic scales on graphs in physics. First, a graph with logarithmic scales enables you to show a much wider range of values, e.g. from 10 to 10 000, which would be impossible to show on a linear scale. Second, a logarithmic graph enables you to find an unknown power in an equation linking two variables, such as n in $y = x^n$.

The logarithm of a number to the base 10 is the power to which 10 must be raised to equal that number. Suppose that $x = 10^y$, we say that y is the logarithm of x to the base 10:

If $x = 10^y$ then $y = \log_{10}x$

For example, $100 = 10^2$ so the logarithm of 100 (to the base 10) is 2.

Suppose you had to plot a graph of the data shown in Table 1.

Table 1

Current/A	Power/W
1	1
10	100
100	10 000
1000	1 000 000
2000	4 000 000
3000	9 000 000

If you used an x-scale that allowed you to plot the first through to the last value on the x-axis (with 1 mm representing 1 amp), you would need rather large graph paper (3 m long!). It would need to be even larger to fit all the y values. Using logarithmic scales on both axes solves this problem.

Using your calculator to find the logarithms of the current and power values, you can add two new columns to the table – as shown in Table 2.

Table 2
Note the headings on the new columns – a logarithm is an index (power) and it has no units, even if the original quantity did have units. 'Log' is used to indicate 'logarithm to the base 10'.

Current/A	Power/W	log(current/A)	log(power/W)
1	1	0.00	0.00
10	100	1.00	2.00
100	10 000	2.00	4.00
1000	1 000 000	3.00	6.00
2000	4 000 000	3.30	6.60
3000	9 000 000	3.48	6.95

This data can now be plotted as a log–log graph.

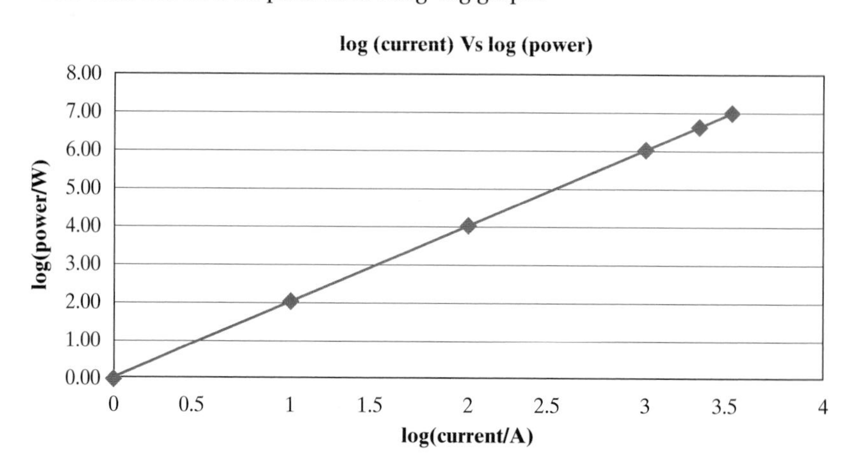

Logarithms transform multiplication to the process of addition:

$$\log_{10}(AB) = \log_{10}A + \log_{10}B \qquad \textbf{Equation 1}$$

This is because when we multiply two powers together we add the indices, e.g. $10^2 \times 10^3 = 10^5$.

Another feature of logarithms is that the process of raising a number to a certain power is transformed to multiplication:

$$\log_{10}(A^n) = n \times \log_{10}A \qquad \textbf{Equation 2}$$

This is because $(10^2)^3 = 10^2 \times 10^2 \times 10^2 = 10^6$, or $10^{2\times3}$.

These two properties of logarithms can be used to plot graphs to find unknown powers. For example, it is known that the force, F, between two bar magnets varies with the distance between them, x, according to the equation:

$$F = Ax^n$$

where A and n are unknown constants.

A graph of F against x would produce an exponential curve, so it is useful to use logarithms. Taking logs of both sides:

$$\log F = \log(Ax^n)$$

Using equation 1, $\log F = \log A + \log x^n$.

Using equation 2, $\log F = \log A + n \log x$.

Compare this to the equation of a straight line graph:

$$y = mx + c$$

If we plot $\log F$ on the y-axis and $\log x$ on the x-axis, the gradient (m) will give the power, n, and the intercept will be equal to $\log A$.

In Physics A-level studies you will often need to use logarithms to the base 10. However, it is sometimes useful to take logarithms to the base e, where e = 2.718. These are called natural logarithms. They are written in the form $\log_e x$, or $\ln x$ ('ln' means natural logarithm). The term $\ln x$ is the inverse function of e^x, so that $\ln e^x = x$.

Natural logarithms are used to plot exponential relationships, such as capacitor discharge and radioactive decay. For example, to find the time constant, CR, from capacitor discharge data giving the potential difference, V, at different times, t:

$$V = V_o e^{\frac{-t}{CR}}$$

$$\frac{V}{V_o} = e^{\frac{-t}{CR}}$$

Taking natural logs of both sides gives:

$$\ln \frac{V}{V_0} = \frac{-t}{CR}$$

so that CR is given by:

$$CR = \frac{-t}{\ln\left(\frac{V}{V_0}\right)}$$

When you are dealing with exponential relationships, it is often useful to plot log–linear graphs, as shown in the example below.

Activity/Bq	Time/s
250	0
226	10
205	20
186	30
168	40
152	50
138	60
125	70

Example

Find the decay constant, and hence the half-life, of protactinium-234 from the data given. Assume that the data has already been corrected for background radiation.

The activity, A, at time t is given by $A = A_0 e^{-\lambda t}$.

Taking logarithms to the base e:

$$\ln A = \ln(A_0\, e^{-\lambda t})$$
$$= \ln A_0 + \ln e^{-\lambda t}$$
$$= \ln A_0 - \lambda t$$

So plotting $\ln A$ on the y-axis against time t on the x-axis gives a straight line with a gradient of $-\lambda$.

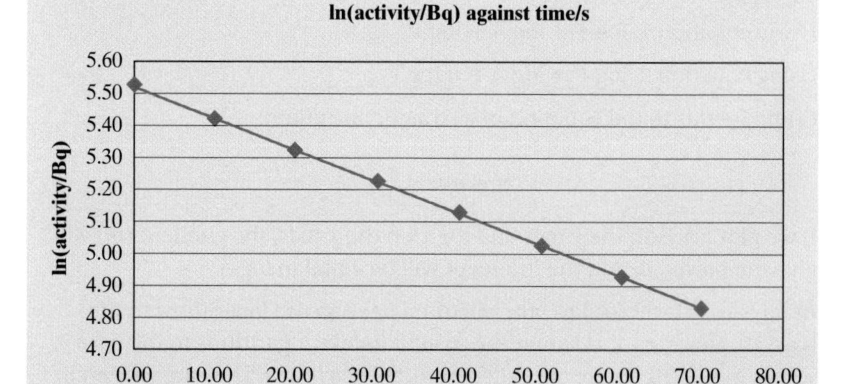

The gradient, $-\lambda$, is $-9.9 \times 10^{-3}\ s^{-1}$.

$$\text{Half-life} = \frac{\ln 2}{\lambda}$$
$$= \frac{0.693}{9.9 \times 10^{-3}\ s^{-1}}$$
$$= 70\ s$$

Data and formulae

FUNDAMENTAL CONSTANTS AND VALUES

Quantity	Symbol	Value	Units
speed of light in vacuo	c	3.00×10^{8}	$m\,s^{-1}$
permeability of free space	μ_0	$4\pi \times 10^{-7}$	$H\,m^{-1}$
permittivity of free space	ε_0	8.85×10^{-12}	$F\,m^{-1}$
charge of electron	e	-1.60×10^{-19}	C
the Planck constant	h	6.63×10^{-34}	J s
gravitational constant	G	6.67×10^{-11}	$N\,m^2\,kg^{-2}$
the Avogadro constant	N_A	6.02×10^{23}	mol^{-1}
molar gas constant	R	8.31	$J\,K^{-1}\,mol^{-1}$
the Boltzmann constant	k	1.38×10^{-23}	$J\,K^{-1}$
the Stefan constant	σ	5.67×10^{-8}	$W\,m^{-2}\,K^{-4}$
the Wien constant	α	2.90×10^{-3}	m K
electron rest mass (equivalent to 5.5×10^{-4} u)	m_e	9.11×10^{-31}	kg
electron charge–mass ratio	e/m_e	1.76×10^{11}	$C\,kg^{-1}$
proton rest mass (equivalent to 1.00728 u)	m_p	1.67×10^{-27}	kg
proton charge–mass ratio	e/m_p	9.58×10^{7}	$C\,kg^{-1}$
neutron rest mass (equivalent to 1.00867 u)	m_n	1.67×10^{-27}	kg
gravitational field strength	g	9.81	$N\,kg^{-1}$
acceleration due to gravity	g	9.81	$m\,s^{-2}$
atomic mass unit (1 u is equivalent to 931.3 MeV)	u	1.661×10^{-27}	kg

ASTRONOMICAL DATA

Body	Mass/kg	Mean radius/m
Sun	1.99×10^{30}	6.96×10^{8}
Earth	5.98×10^{24}	6.37×10^{6}

GEOMETRICAL EQUATIONS

arc length $= r\theta$

circumference of circle $= 2\pi r$

area of circle $= \pi r^2$

area of cylinder $= 2\pi rh$

volume of cylinder $= \pi r^2 h$

area of sphere $= 4\pi r^2$

volume of sphere $= \frac{4}{3}\pi r^3$

MOMENTUM

force $F = \frac{\Delta(mv)}{\Delta t}$

impulse $F\,\Delta t = \Delta(mv)$

CIRCULAR MOTION

angular velocity $\omega = \frac{v}{r}$

$\omega = 2\pi f$

centripetal acceleration $a = \frac{v^2}{r} = \omega^2 r$

centripetal force $F = \frac{mv^2}{r} = m\omega^2 r$

OSCILLATIONS

acceleration $a = -(2\pi f)^2 x$

displacement $x = A\cos(2\pi ft)$

speed $v = \pm 2\pi f\sqrt{A^2 - x^2}$

maximum speed $v_{max} = 2\pi fA$

maximum acceleration $a_{max} = (2\pi f)^2 A$

for a mass-spring system $T = 2\pi\sqrt{\frac{m}{k}}$

for a simple pendulum $T = 2\pi\sqrt{\frac{l}{g}}$

GRAVITATIONAL FIELDS

force between two masses $F = -\frac{Gm_1 m_2}{r^2}$

gravitational field strength $g = \frac{F}{m}$

magnitude of gravitational field strength in radial field $g = \frac{GM}{r^2}$

gravitational potential $\Delta W = m\,\Delta V$

$V = -\frac{GM}{r}$

$g = -\frac{\Delta V}{\Delta r}$

ELECTRIC FIELDS AND CAPACITORS

force between two point charges $F = \frac{1}{4\pi\varepsilon_0}\frac{Q_1 Q_2}{r^2}$

force on a charge $F = EQ$

field strength for a uniform field $E = \frac{V}{d}$

field strength for a radial field $E = \frac{Q}{4\pi\varepsilon_0 r^2}$

electric potential $\Delta W = Q\,\Delta V$

$V = \frac{1}{4\pi\varepsilon_0}\frac{Q}{r}$

capacitance $C = \frac{Q}{V}$

decay of charge $Q = Q_0 e^{-t/RC}$

time constant RC

capacitor energy stored $E = \frac{1}{2}QV = \frac{1}{2}CV^2 = \frac{1}{2}\frac{Q^2}{C}$

MAGNETIC FIELDS

force on a current $F = BIl$

force on a moving charge $F = BQv$

magnetic flux $\Phi = BA$

magnetic flux linkage $\Phi = BAN$

induced emf $\varepsilon = N\frac{\Delta\Phi}{\Delta t}$

emf induced in a rotating coil $N\Phi = BAN\cos\theta$

$\varepsilon = BAN\omega\sin\omega t$

transformer equations $\frac{N_s}{N_p} = \frac{V_s}{V_p}$

efficiency $= \frac{I_s V_s}{I_p V_p}$

RADIOACTIVITY AND NUCLEAR PHYSICS

the inverse square law for γ radiation $I = \frac{k}{x^2}$

radioactive decay $\frac{\Delta N}{\Delta t} = -\lambda N$

$N = N_0 e^{-\lambda t}$

activity $A = \lambda N$

half life $T_{1/2} = \frac{\ln 2}{\lambda}$

nuclear radius $R = r_0 A^{1/3}$

energy–mass equation $E = mc^2$

GASES AND THERMAL PHYSICS

gas law $pV = nRT$

$pV = NkT$

kinetic theory model $pV = \frac{1}{3}Nm\left(c_{rms}\right)^2$

kinetic energy of gas molecule $\frac{1}{2}m\left(c_{rms}\right)^2 = \frac{3}{2}kT = \frac{3RT}{2N_A}$

energy to change temperature $Q = mc\,\Delta T$

energy to change state $Q = ml$

ASTROPHYSICS

1 astrophysics unit = 1.50×10^{11} m

1 light year = 9.45×10^{15} m

1 parsec = 206265 AU = 3.08×10^{16} m = 3.261 yr

Hubble constant H = 65 km s^{-1} Mpc^{-1}

lens equation $\frac{1}{f} = \frac{1}{u} + \frac{1}{v}$

$$M = \frac{\text{angle subtended by image at eye}}{\text{angle subtended by object at unaided eye}}$$

in normal adjustment $M = \frac{f_0}{f_e}$

resolving power $\theta \approx \frac{\lambda}{D}$

magnitude equation $m - M = 5 \log \frac{d}{10}$

Wien's law $\lambda_{max} T = 0.0029$ m K

Hubble law $v = Hd$

Stefan's law $P = \sigma A T^4$

Doppler shift for $v << c$ $z = \frac{\Delta f}{f} = -\frac{\Delta\lambda}{\lambda} = \frac{v}{c}$

Schwarzschild radius $R_s = \frac{2GM}{c^2}$

MEDICAL PHYSICS

lens equations $P = \frac{1}{f}$

$m = \frac{v}{u}$

$\frac{1}{f} = \frac{1}{u} + \frac{1}{v}$

intensity level $\text{intensity level} = 10 \log \frac{I}{I_0}$

absorption $I = I_0 e^{-\mu x}$

$\mu_m = \frac{\mu}{\rho}$

APPLIED PHYSICS

moment of inertia $I = \Sigma mr^2$

angular kinetic energy $E_k = \frac{1}{2} I\omega^2$

equations of angular motion $\omega_2 = \omega_1 + \alpha t$

$\omega_1^2 = \omega_1^2 + 2\alpha\theta$

$\theta = \omega_1 t + \frac{1}{2}\alpha t^2$

torque $T = I\alpha$

angular momentum $\text{angular momentum} = I\omega$

work done $W = T\theta$

power $P = T\omega$

thermodynamics $Q = \Delta U + W$

$W = p\Delta V$

adiabatic change $pV^\gamma = \text{constant}$

isothermal change $pV = \text{constant}$

heat engines

efficiency = $\frac{W}{Q_{in}} = \frac{Q_{in} - Q_{out}}{Q_{in}}$

maximum efficiency = $\frac{T_H - T_C}{T_H}$

work done per cycle = area of loop

input power = calorific value × fuel flow rate

indicated power = (area of p-V loop) × (no. of cycles per second) × number of cylinders

output of brake power $P = T\omega$

friction power = indicated power – brake power

heat pumps and refrigerators

refrigerator: $COP_{ref} = \frac{Q_{out}}{W} = \frac{Q_{out}}{Q_{in} - Q_{out}}$

heat pump: $COP_{hp} = \frac{Q_{in}}{W} = \frac{Q_{in}}{Q_{in} - Q_{out}}$

TURNING POINTS IN PHYSICS

electrons in fields $F = \frac{eV_p}{d}$

$F = Bev$

$r = \frac{mv}{Be}$

$\frac{1}{2}mv^2 = eV$

$\frac{QV}{d} = mg$

$F = 6\pi\eta r v$

wave particle duality $c = \frac{1}{\sqrt{\mu_0 \varepsilon_0}}$

$\lambda = \frac{h}{p} = \frac{h}{\sqrt{2meV}}$

special relativity $E = mc^2 = \frac{m_0 c^2}{\left(1 - \frac{v^2}{c^2}\right)^{\frac{1}{2}}}$

$l = l_0 \left(1 - \frac{v^2}{c^2}\right)^{\frac{1}{2}}$ $t = t_0 \left(1 - \frac{v^2}{c^2}\right)^{-\frac{1}{2}}$

Practice exam-style questions

Section A

For each question there are four responses. Select the response which you think is the most appropriate answer to a question. One mark for each correct answer.

1 A footballer kicks a stationary ball of mass 0.40 kg. The contact lasts for 0.025 s and the ball leaves his foot at 20 m s^{-1}. What is the average force on the ball?

A 320 N

B 32 N

C 0.2 N

D 80 N

2 The graph shows how the force on a bicycle changes with time.

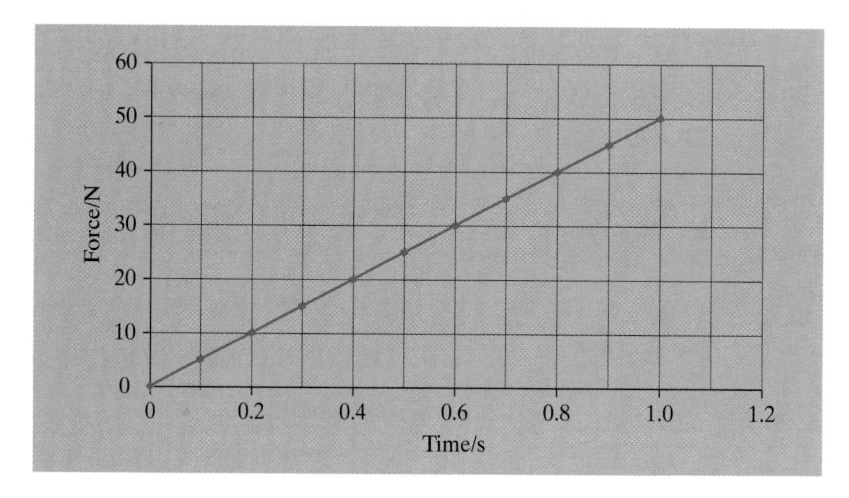

The change in momentum of the bicycle over the 1 s period is

A 50 kg m s^{-1}

B 25 kg m s^{-1}

C 50 N s^{-1}

D 25 N s^{-1}

3 A spin drier of radius 20 cm spins at 1200 rpm (revolutions per minute). The speed of a point on the edge of the drum is

A 24 000 m s^{-1}

B 240 m s^{-1}

C 0.8 m s^{-1}

D 25 m s^{-1}

4 A satellite in a circular orbit above the Earth's atmosphere will stay in orbit, without transferring any energy. This is because

A The resultant force is along a tangent to the circle.

B The centripetal force is parallel to the velocity.

C The centrifugal force is parallel to the velocity.

D The centripetal force is perpendicular to the velocity.

5 A car of mass 1200 kg is travelling at 25 m s^{-1}. If the maximum frictional force between the car and the road is 5000 N, what is the radius of the tightest corner that the car can safely drive around?

A 150 m

B 1500 m

C 6 m

D 520 m

6 The diagram shows a pendulum swinging.

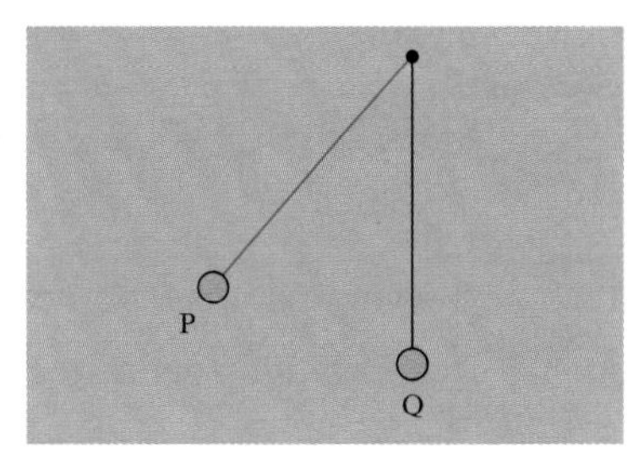

Choose the row in the table which correctly shows where the maximum velocity and maximum acceleration occur.

	Maximum velocity	Maximum acceleration
A	Q	P
B	P	Q
C	P	P
D	Q	Q

7 A body moves with simple harmonic motion of period 2.00 s. If the displacement of the body from the equilibrium position is 0.50 m at the start of the motion, what is the displacement of the body after 0.30 s?

A 0.50 m

B 0.59 m

C 0.15 m

D 0.29 m

8 A spring is made to vibrate by attaching it to a vibration generator. The vibration generator can be driven at different frequencies by the signal generator. The oscillations of the vibration generator are transmitted to the spring by a string which passes over a pulley.

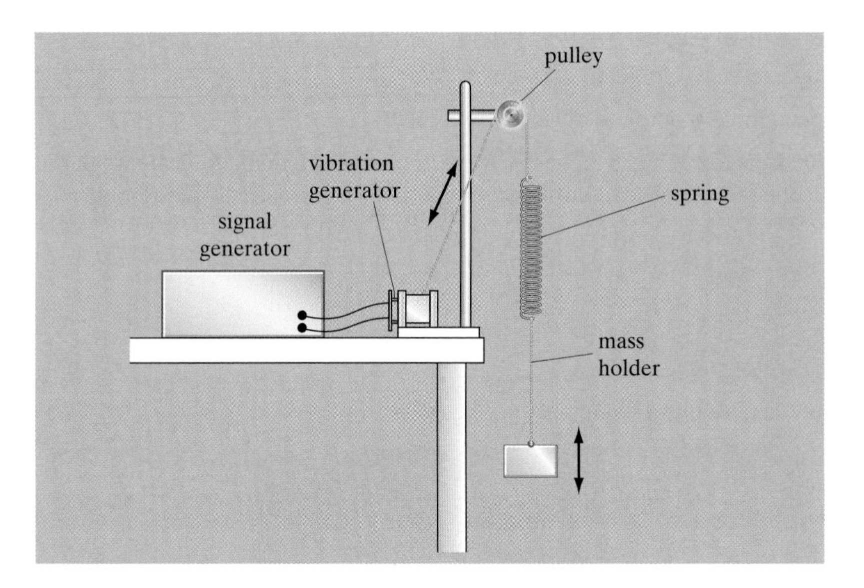

Which one of these statements about the forced vibrations is *not* true?

A Resonance occurs when the frequency of the vibration generator matches the natural frequency of the mass and springs.

B At very low frequencies the mass vibrates at the same frequency as the vibration generator.

C When the vibration generator oscillates at very high frequencies, the mass hardly moves.

D At resonance the oscillations of the vibration generator are in phase with the mass-spring oscillations.

9 Which one of these statements about simple harmonic motion is true?

A The damping force is always in the opposite direction to the velocity.

B The acceleration is always in the opposite direction to the velocity.

C The velocity is proportional to the displacement from equilibrium.

D The acceleration is always in the same direction as the velocity.

10 A 2 kg mass is oscillating on a spring of spring constant *k*. If the mass oscillates with a period of 0.5 s, the value of *k* is

A 158 N m^{-1}

B 25 N m^{-1}

C 316 N m^{-1}

D 31.6 N m^{-1}

11 Magnetic flux density is measured in tesla, T. One tesla is equivalent to

A 1 N A m^{-1}

B 1 N A^{-1} m^{-1}

C 1 Wb m^{-1}

D 1 Wb m^2

12 The diagram shows an electron entering a magnetic field. Which arrow shows the direction of the force on the electron as it enters the field?

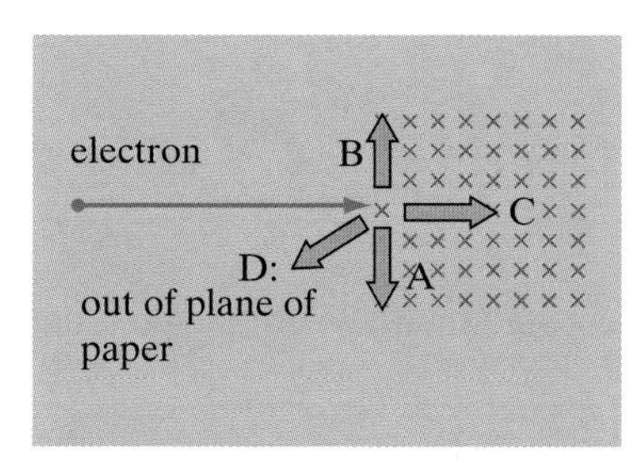

13 The gravitational field strength at the surface of the Earth is 9.81 N kg^{-1}. The planet Zob has the same average density as Earth, but has twice the diameter. What will the gravitational field strength be on the surface of Zob?

A 9.81 N kg^{-1}

B 19.6 N kg^{-1}

C 39.2 N kg^{-1}

D 4.90 N kg^{-1}

14 Which graph shows how the gravitational potential changes with distance, r, from the centre of the Earth?

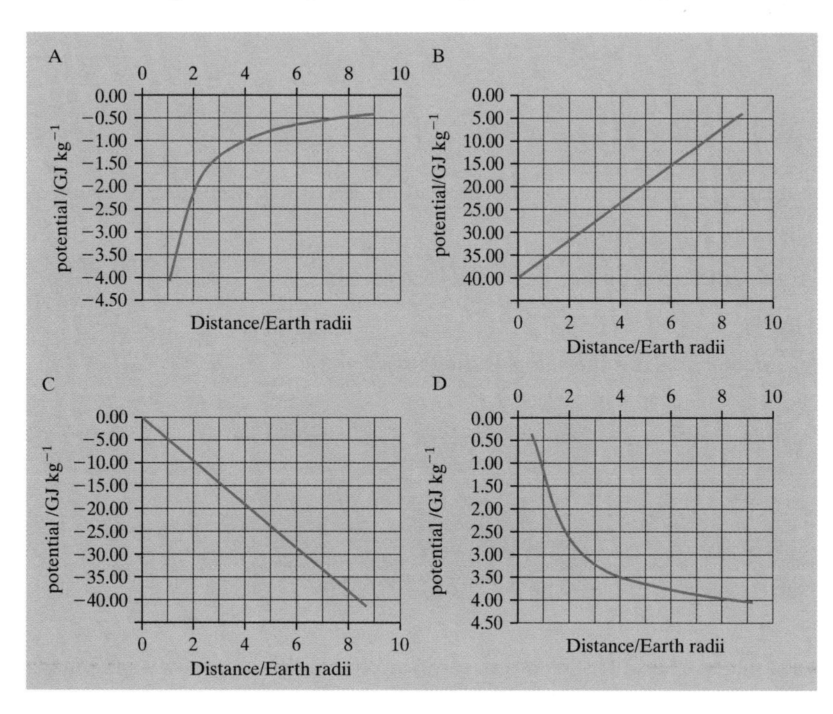

15 An electron enters the uniform electric field between two charged plates. The plates are 3 cm apart and there is a potential difference across them of 5 kV. The acceleration of the electron will be

A 2.9×10^{13} m s^{-2}

B 2.9×10^{14} m s^{-2}

C 2.9×10^{15} m s^{-2}

D 2.9×10^{16} m s^{-2}

16 Which row in the table has *two* correct statements about charged particles in electric and magnetic fields?

	Electric fields	Magnetic fields
A	Affects moving charges only	Changes the speed of charged particles
B	Changes the speed of charged particles	Affects moving charges only
C	Causes a parabolic path	Changes the speed of charged particles
D	Affects stationary charges only	Causes a circular path

17 JJ Thomson found the charge–mass ratio (e/m) of an electron by passing an electron beam through a magnetic field, B, and an electric field, E, at the same time. By balancing the effects of the two fields he managed to keep the electron beam straight (undeflected). If the electrons in the beam were moving with a velocity v, which one of these equations had to be satisfied to keep the beam straight?

A $eE = \frac{1}{2}mv^2$

B $E = Bv$

C $E = Bev$

D $(e/m) = B/E$

18 A satellite is moving in a geosynchronous orbit around the Earth. Which of the following statements is *false*?

A The satellite is in an equatorial orbit.

B The radius of the orbit depends on the mass of the satellite.

C The orbital period is 24 hours.

D The satellite's angular speed is the same as the Earth's.

19

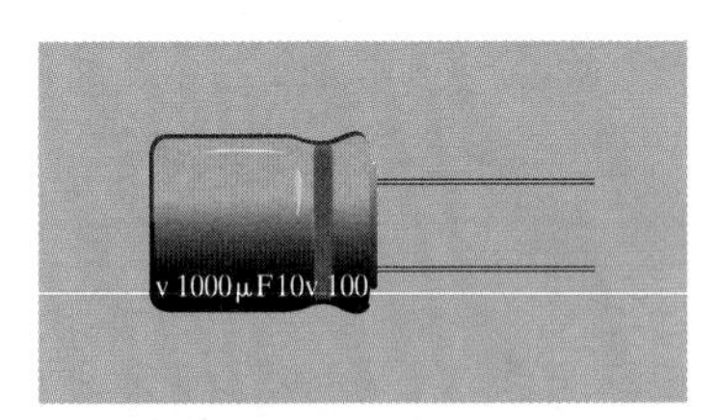

The diagram shows a capacitor. Identify the correct row from the table, which shows the maximum charge and the maximum energy that can be stored by the capacitor.

	Maximum charge stored /mC	Maximum energy stored /J
A	0.01	0.1
B	10	0.1
C	0.01	0.05
D	10	0.05

20 A capacitor of capacitance C is charged to a voltage V, and then discharged through a resistance R. The time taken for the voltage to drop to $\frac{1}{2}V$ is given by

A CR

B $2CR$

C $CR \ln 2$

D $\frac{1}{2}CR$

21 A coil of area A is inclined so that the normal to plane of the coil is at an angle α to a magnetic field of flux density B. If the coil has N turns, the flux linkage through the coil is given by

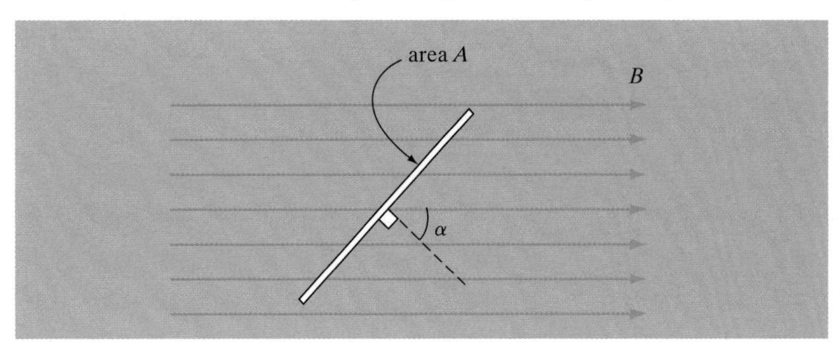

A $BA \cos \alpha$

B $BAN \sin \alpha$

C $BAN \cos \alpha$

D BAN

22 A step-down transformer is used to provide a 12 V supply when plugged into the mains supply (230 V) in a laboratory. The transformer is 80% efficient and the current through the primary coil is 0.5 A. Which row in the table is correct?

	Current through the secondary /A	**Turns on primary**	**Turns on secondary**
A	3.8	26	500
B	0.2	26	500
C	3.8	500	26
D	7.6	500	26

23 Transformers are used in the National grid. Which of these statements is *not* true?

A Some electrical energy is always dissipated as heat in a transformer.

B Step-down transformers are used near houses to reduce the voltage to safer levels.

C Alternating current is used as transformers do not work with d.c.

D Electrical power is transmitted at high current to reduce power losses in the transmission cables.

24 An aircraft is flying east on a horizontal flight-path over the UK, where the Earth's magnetic field has a flux density of 45 μT inclined at an angle of 65° below the horizontal. The aircraft has a wing span of 30 m and is flying at 150 m s^{-1}. What is the e.m.f. induced between the tips of the aircraft wings?

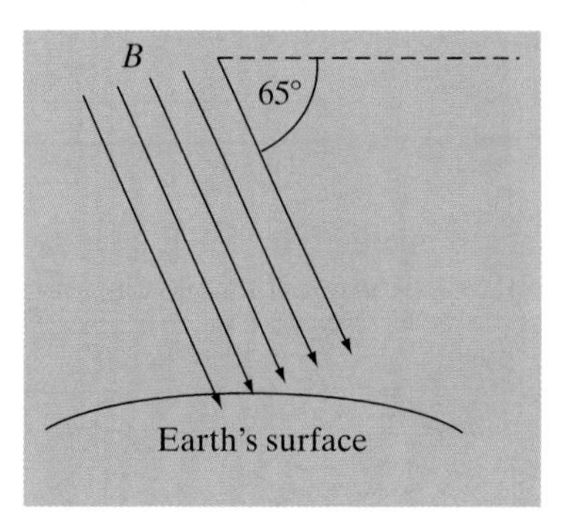

A 0.18 V

B 0.08 V

C 180 V

D 80 V

25 A generator consists of a rectangular coil of area 0.001 m^2, with 500 turns, spinning at 50 revolutions per second in a field of flux density 0.2 T. An oscilloscope is used to examine the e.m.f. induced. The time base is set to 5 ms per division and the *Y*-gain is set to 10 V per division. Which screen shows the oscilloscope output correctly?

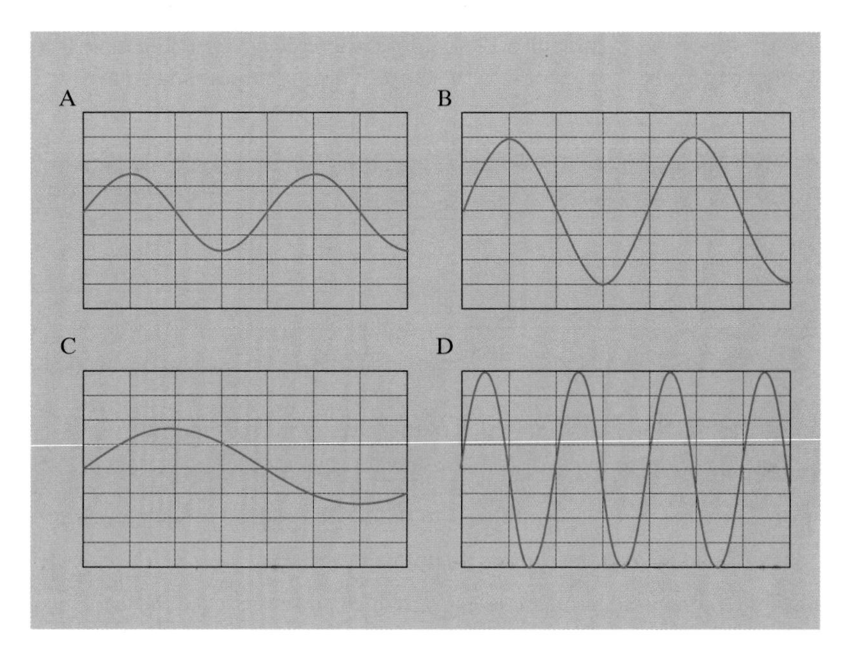

Section B

1 A nail of mass 30 g is being driven into a block of wood. Its initial speed is 30 m s^{-1} and it is brought to rest in the wood in 5 ms.

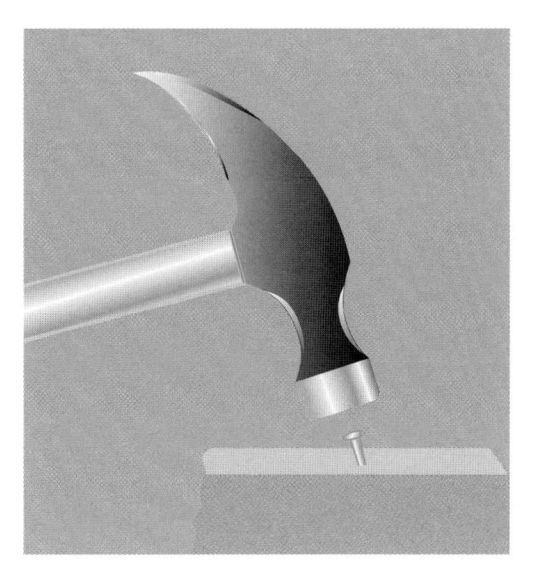

(a) **(i)** Calculate the impulse acting on the nail.

____

____

____ 2 marks

(ii) Calculate the average force acting on the nail.

____

____ 2 marks

(b) Use the idea of impulse to explain how the crumple zone of a car helps to protect passengers during a collision.

____

____

____

____ 4 marks

Total Marks: 8

2 A cathode ray tube uses a high voltage anode to accelerate a beam of electrons across a vacuum tube. The potential difference between the cathode and the anode is 2000 V.

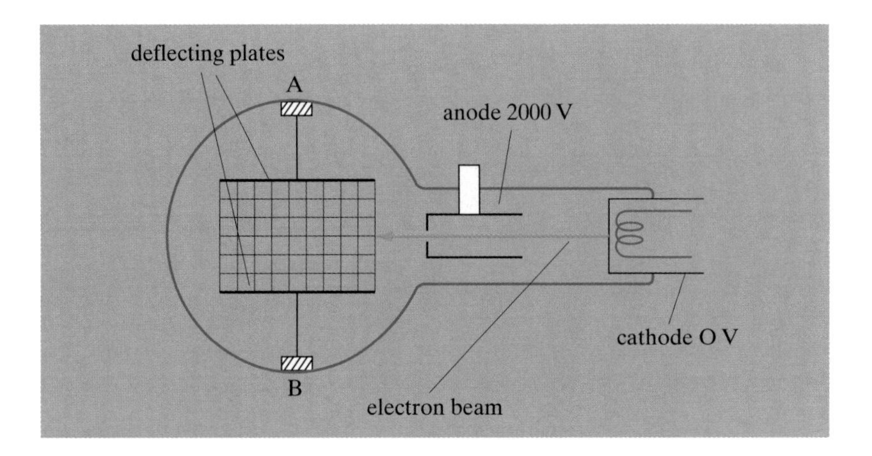

(a) Calculate the energy gained by an electron as it accelerates from the cathode to the anode. Give your answer in joules.

2 marks

(b) Show that the velocity of an electron as it reaches the anode is 2.7×10^7 m s^{-1}.

2 marks

(c) The deflecting plates are 5 cm apart. A potential of +3 kV is applied to the top plate. The bottom plate is connected to 0 V.

(i) Describe the path of the electron as it travels between the plates.

2 marks

(ii) Calculate the electric field strength between the deflection plates.

2 marks

(iii) Calculate the acceleration of the electron as it moves between the deflection plates.

2 marks

Total Marks: 10

3 A child on a playground swing is pulled back and released, so that she oscillates with a natural frequency of 0.4 Hz.

(a) Explain what is meant by the natural frequency of a system.

2 marks

(b) If the child and swing behave like a simple pendulum, calculate the length of the swing.

2 marks

(c) Calculate the displacement after 1 second has elapsed. (You may neglect the effect of damping.)

2 marks

(d) The child is pulled back so that the initial displacement is 0.5 m, and then released. The oscillations are damped.

(i) Explain what is meant by 'damped' oscillations.

2 marks

(ii) Sketch a graph showing three full oscillations. 4 marks

Total Marks: 12

4 Particle accelerators use strong magnetic fields to steer charged particles around bends. In one accelerator a proton travelling at $0.3c$ enters a magnetic field of flux density 10 T.

(a) Calculate the force on the proton.

2 marks

(b) The proton moves in a circular path. Calculate the radius of the circular path.

2 marks

(c) Calculate the time taken for the proton to complete one complete circle.

2 marks

(d) Explain why magnetic fields cannot be used to increase the proton's speed.

____

____ 2 marks

Total Marks: 8

5 A bar magnet is dropped through a coil of wire that is connected to an oscilloscope. The output of the oscilloscope is shown below.

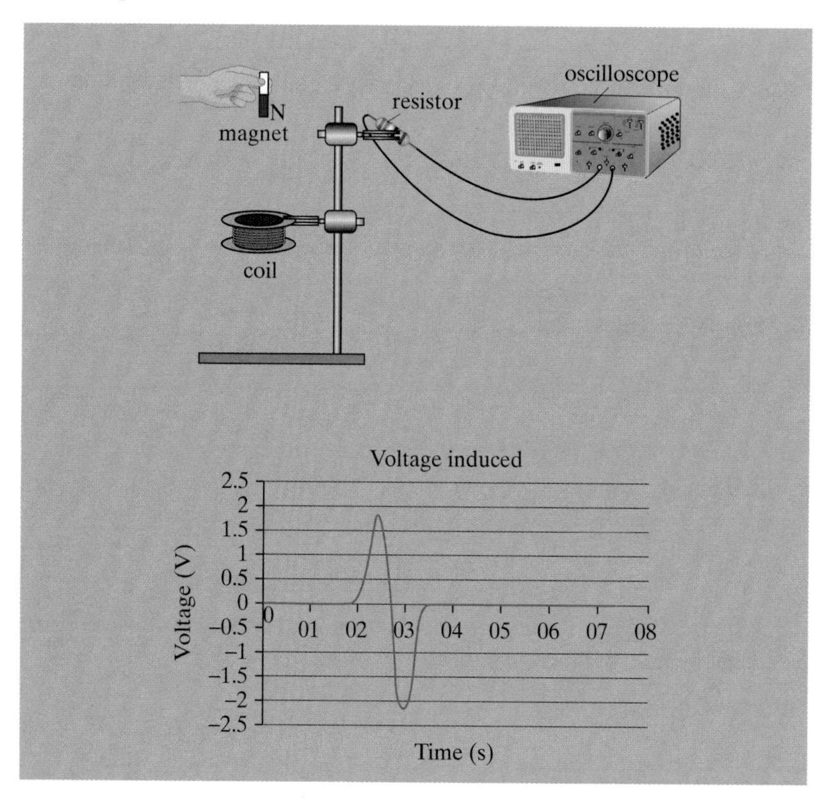

(a) Explain why a voltage is induced in the coil.

____

____

____ 2 marks

(b) State and explain what would happen to the oscilloscope display if the magnet was dropped from a greater height.

____

____

____ 2 marks

(c) How would the trace change if the magnet was turned, so that the south pole entered the coil first?

1 mark

Total Marks: 5

6

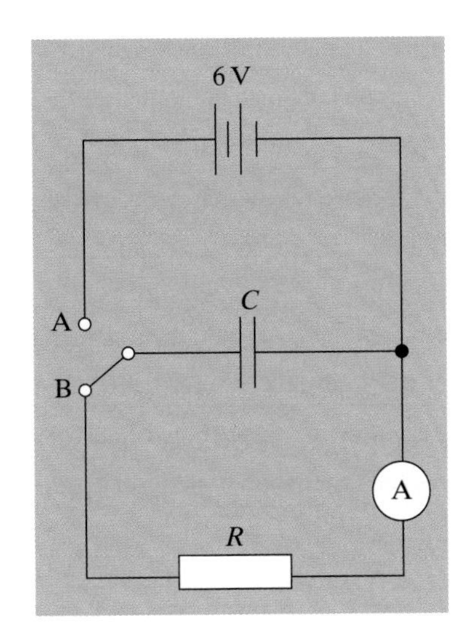

A capacitor is initially connected to the 6 V battery by moving the switch to A. At time $t = 0$, the switch is moved to B and the capacitor is discharged through the resistor. The capacitor has a capacitance of 100 mF, and the resistor has a resistance of 1 kΩ.

(a) Explain what is meant by the 'time constant' of this circuit.

1 mark

(b) Calculate the time constant for this circuit.

2 marks

(c) What will the initial current reading on the ammeter be when the switch is moved from A to B?

1 mark

(d) How long will it take for the ammeter reading to drop to 1 mA?

3 marks

Total Marks: 7

Answers, explanations, hints and tips

Section A

1 mark for each correct answer.

Question	Answer	Comment
1	A	Force = change in momentum/time = $0.40 \times 20/0.025 = 320$ N
2	B	Area under F–t graph is change in momentum = 25 kg m s^{-1} (or N s but not N s^{-1}).
3	D	$f = 1200/60 = 20$ Hz; $v = 2\pi fr = 25(.1)$ m s^{-1} ($r = 0.2$ m)
4	D	For circular motion, the resultant force is towards the centre and at right angles to the velocity, so no work is done.
5	A	$F = mv^2/r$; so $r = mv^2/F = 1200 \times 25^2/5000 = 150$ m
6	A	Maximum acceleration occurs at maximum displacement.
7	D	$x = A \cos \omega t = 0.5 \times \cos (2 \times \pi \times f \times 0.3) = 0.29$ m. Remember $f = 1/T$ and use radians.
8	D	There is a phase difference of $\pi/2$ between driving system and driven system at resonance.
9	A	Resistive forces such as friction or drag act in opposition the velocity.
10	C	$T = 2\pi \sqrt{(m/k)}$; so $k = 4\pi^2 m/T^2 = 315.8$ N m^{-1}
11	B	From $F = BIl$ (1 T is also equivalent to 1 Wb m^{-2}, but not Wb m^2).
12	A	Use Fleming's left-hand rule: first finger (field) into paper, second finger (current) in opposite direction to the electron's motion (since the electron is negatively charged).
13	B	Twice the diameter means 8 times the mass (since mass depends on volume and $V = \frac{4}{3}\pi R^3$; $g_{Earth} = GM/R^2$; $g_{Zob} = G \times 8M/4R^2 = 2 \times g_{Earth}$
14	A	Potential is defined as zero at infinity and gets more negative nearer Earth, and $V \propto 1/r$.
15	D	Field strength is $V/d = 5000/0.03 = 167$ N C^{-1}; so force is $167 \times 1.6 \times 10^{-19} = 2.67 \times 10^{-14}$ N, and $a = F/m = 2.67 \times 10^{-14}/9.11 \times 10^{-31}$ N $kg^{-1} = 2.93 \times 10^{16}$ m s^{-2}
16	B	Magnetic fields only affect moving charges and change their direction, but not speed.
17	B	Balancing the forces due to the magnetic and electric fields, $Bev = eE$, so $Bv = E$.
18	B	The radius of any orbit depends on speed, but not mass.
19	D	$C = 1000 \times 10^{-6}$ F and maximum $V = 10$ V; $Q = CV = 0.01$ C and $E = \frac{1}{2}QV = 0.05$ J
20	C	$V = V_0 \exp(-t/CR)$; so $\ln(V_0/V) = t/CR$, and $V_0/V = 2$ in this case.
21	C	The flux linkage is AN times the component of B perpendicular to A.
22	D	Step-down the voltage by a factor $12/230 = 1/19$, so current goes up by 19 times to 9.5 A, but transformer only 80% efficient so $I = 7.6$ A. Need a ratio of 19 for turns, so 500 primary and 26 secondary is correct.
23	D	Power is transmitted at high voltage, not high current.
24	A	Component B vertically is $B_v = 45 \times 10^{-6} \cos 25 = 40.8 \times 10^{-6}$ T. Area swept through in 1 second = wing span × velocity, so induced e.m.f. = $B_v L v = 0.18$ V
25	B	Peak voltage is $BAN\omega = 0.2 \times 0.001 \times 500 \times 2 \times \pi \times 50 = 31$ V, which is about 3 divisions (3×10); period = $1/f = 0.02$ s, which is 4 divisions (4×0.005).

Section B

Question	Answer		Marks
1 (a)(i)	Impulse = force × time = change in momentum = 0.03 × 30 = 0.9 N s (1 mark for correct unit.)	 (2)	
(ii)	Force = impulse/time = 0.9/0.005 = 180 N	(1) (1)	4
1 (b)	Need to reduce the force on the passengers. The impulse is fixed/change of momentum is fixed by mass and change in speed. Crumple zone increases t. Impulse is Ft; so if t increases, F decreases.	(1) (1) (1) (1)	4
			Total 8
2 (a)(i)	Energy = $eV = 1.6 \times 10^{-19} \times 2000$ $= 3.2 \times 10^{-16}$ J	(1) (1)	2
2 (b)	Energy = $\frac{1}{2}mv^2$; so $v = \sqrt{(2E/m)}$ $= 2.65 \times 10^7$ m s^{-1} (In 'show that' questions, always give the next significant figure to prove you actually did the calculation.)	(1) (1)	2
2 (c)(i)	Curving upwards in a parabolic (not circular) path.	(1) (1)	
(ii)	Field strength is 3000/0.05 = 60 kV m^{-1} or kN C^{-1} (1 mark for correct unit.)	(2)	
(iii)	Acceleration = $F/m = eE/m = 1.6 \times 10^{-19} \times 60\,000/9.11 \times 10^{-31}$ $= 1.1 \times 10^{16}$ m s^{-2}	(1) (1)	6
			Total 10
3 (a)	The number of free/unforced oscillations per second. Free oscillations occur when an object is disturbed from equilibrium and released.	(1) (1)	2
3 (b)	$T = 2\pi\sqrt{(l/g)}$; so $l = T^2g/4\pi^2$ = 1.6 m (T = 1/0.4 = 2.5 s)	(1) (1)	2
3 (c)	$x = A\cos\omega t = 0.5\cos(2\pi ft) = 0.5\cos(2\pi \times 0.4 \times 1)$ = −0.40 m (Must have −ve sign for second mark.)	(1) (1)	2
3 (d)(i)	Damped oscillations occur when energy is transferred from the oscillating system/work is done against resistive forces. The amplitude of the oscillations decreases with time.	(1) (1)	
(ii)	x and y axes labelled and suitable scales. Period = 0.4 s for all 3 cycles. Initial amplitude = 0.5 m, subsequent amplitude gets smaller.	(1) (1) (1) (1)	6
			Total 12

Question	Answer		Marks
4 (a)	$F = Bqv = 10 \times 1.6 \times 10^{-19} \times 0.3 \times 3 \times 10^8$ $= 1.44 \times 10^{-10}$ N	(1) (1)	 2
4 (b)	$F = mv^2/r$; so $r = mv^2/F = 1.67 \times 10^{-27} \times (0.3 \times 3 \times 10^8)^2/1.44 \times 10^{-10}$ $= 0.094$ m	(1) (1)	 2
4 (c)	Time for one circle = dist/speed = $2\pi r/v$ $= 6.6 \times 10^{-9}$ s	(1) (1)	 2
4 (d)	Force always acts at right angles to velocity. Does no work/just changes direction.	(1) (1)	 2
			Total 8
5 (a)	As the magnet drops through the coil, the flux linkage changes. Faraday's law states that e.m.f. is equal to the rate of change of flux linked.	(1) (1)	 2
5 (b)	Induced voltage would be greater/trace has greater amplitude. Magnet moving quicker, so rate of change of flux is greater/less time to pass through coil.	(1) (1)	 2
5 (c)	The trace would be inverted/go negative first.	(1)	1
			Total 5
6 (a)	The time constant is the time taken for the voltage/current/charge to drop to 1/e of its initial value.	(1)	1
6 (b)	$CR = 100 \times 10^{-3} \times 1 \times 10^3$ $= 100$ s	(1) (1)	 2
6 (c)	$I = V/R = 6/1000 = 0.006$ A or 6 mA	(1)	1
6 (d)	$I = I_0 \exp(-t/CR)$; so $I/I_0 = 1/6$ and $I_0/I = 6$ $t = CR \ln 6$ $= 180$ s	(1) (1) (1)	 3
			Total 7

Glossary

amplitude	the largest displacement from equilibrium of an oscillating object, or the maximum height of a wave
angular speed (ω)	in rotation, the angle turned through per second; unit radians per second, rad s^{-1}
capacitance	the charge stored per unit potential difference applied across a capacitor; unit farad, F
capacitor	a device that stores electric charge, and therefore electrical energy
centripetal acceleration	an acceleration towards the centre of a circular path
centripetal force	a force causing an object to move in a circular path; it acts towards the centre of the circle
conservation of linear momentum	a fundamental conservation law in physics: the total momentum of any system is constant, in the absence of external forces
Coulomb's law	the law describing the electrostatic force between two charges: the force is proportional to the product of the charges and inversely proportional to the square of the distance between them
damped oscillation	oscillation for which the amplitude decreases with time as work is done against resistive forces
dielectric	the insulating material between the plates of a capacitor
displacement (s)	a vector describing the difference in position of two points
driving frequency	the frequency of an external varying force which is causing oscillations
elastic collision	a collision in which the total kinetic energy is conserved, i.e. the sum of the kinetic energies before and after the collision is equal
electric field strength	the force on a unit charge: $E = F/Q$
electric potential (V)	property of a point in an electric field: the work done in bringing a unit positive charge from infinity to that point; unit joules per coulomb, or volts
electron volt (eV)	a unit of energy equal to the energy transferred when an electron moves through a potential difference of 1 volt; 1 eV = 1.6×10^{-19} J
escape velocity	the minimum initial velocity needed at the surface of a planet or star if an object is to escape its gravitational pull; on Earth an object thrown upwards at 11 km s^{-1} would totally escape the Earth's gravity and never fall down again
exponential decay	when a quantity reduces in magnitude by a certain factor, e.g. half, in a constant time period it is said to decay exponentially
farad (F)	the unit of capacitance; a 1 F capacitor would store 1 C of charge for every 1 V of potential difference applied across it
Faraday's law (of electromagnetic induction)	when there is relative movement between a conductor and a magnetic field, the e.m.f. induced in a conductor is equal to the rate of change of magnetic flux linkage
forced vibration	an oscillation caused by an external varying force

frequency	the number of oscillations or waves in one second; unit hertz, Hz.
fundamental mode	the lowest frequency of a resonating standing wave system
geostationary satellite	a satellite in an equatorial orbit over a fixed point on the Earth's surface; its orbital period is 24 hours
gravitational field strength (g)	the force on a unit mass at a point in a gravitational field; unit N kg^{-1}
gravitational potential (V)	property of a point in a gravitational field: the work done in bringing a unit mass from infinity to that point; unit joules per kilogram
impulse	the magnitude of a force (F) multiplied by the time (t) for which it acts: impulse $= Ft$
induced e.m.f.	a potential difference across a conductor in a complete circuit, caused by a change in the magnetic flux around the conductor
inelastic collision	a collision in which kinetic energy is not conserved
Lenz's law	the direction of an induced e.m.f. is such that it opposes the change of magnetic flux that caused it
magnetic flux (Φ)	the magnetic flux through an area A is equal to (flux density perpendicular to A) $\times A$; unit weber, Wb
magnetic flux density (B)	a measure of the strength of a magnetic field; the flux density is equal to the magnitude of the force on a unit length of current-carrying wire per unit current; unit tesla, T
magnetic flux linkage	the total magnetic flux linkage through a coil of area A and number of turns N is $N\Phi$ or NAB
momentum (p)	property of a moving object equal to its mass (m) multiplied by its velocity (v), $p = mv$; a vector quantity, unit kg m s^{-1}
motor effect	a wire carrying a current in a magnetic field experiences a force; this is known as the motor effect
natural frequency	the frequency at which a vibrating object undergoes free vibrations; for example, the frequency at which a tuning fork will oscillate when struck
Newton's law of gravitation	the gravitational force between two point masses is proportional to the product of the masses, and inversely proportional to the square of the distance between them
oscillation	a repeating to-and-fro motion, such as that observed in a swinging pendulum or vibrating guitar string
period (T)	the time taken to one complete oscillation, or one full rotation
periodic motion	a repeating pattern of motion, such as rotation or oscillation
permittivity (of free space)	the permittivity of a medium is a measure of the material's ability to resist the formation of an electric field within it; using a dielectric with a permittivity greater than that of free space (a vacuum) increases the capacitance of a capacitor
phase difference	the difference in position in the cycle of two oscillating systems or waves, expressed in degrees or radians

potential difference	the (electric) potential difference between two points is the energy transferred per unit charge moving between the points
radian	a unit used to measure angle; 1 radian is the angle subtended at the centre of a circle by an arc whose length is equal to the radius; 2π radians = 360°
resonance	large amplitude vibrations caused when the driving frequency matches the natural frequency of a system
restoring force	a force that acts so as to return an object to its equilibrium position
rotational frequency	the number of rotations per second; unit hertz, Hz (or sometimes revolutions per minute, rpm, but this is not an SI unit)
scalar	a physical quantity that is fully specified by its magnitude (size); it has no direction associated with it
simple harmonic motion (SHM)	periodic motion such that the restoring force is proportional to the displacement from equilibrium and acts in the opposite direction
stationary wave	a wave that does not transfer energy in the direction of wave travel
tesla (T)	unit of magnetic flux density; 1 T is the magnetic flux density when 1 m of wire carrying 1 A of current at right angles to a magnetic field experiences a force of 1 N
time constant	time taken for the charge on (or voltage across or current flowing off) a discharging capacitor to drop to $1/e$ (≈ 0.37) of its original value; it is equal to CR, where C is the capacitance and R is the total resistance of the circuit; unit seconds
transformer	a device that uses electromagnetic induction to change the voltage (and current) of an a.c. signal
transformer equation	$N_s/N_p = V_s/V_p$
vector	a physical quantity that is specified by its magnitude (size) and its direction
weber (Wb)	the unit of magnetic flux; 1 Wb = 1 T m^2

Index